Stage to Studio

STUDIES IN INDUSTRY AND SOCIETY

Philip B. Scranton, Series Editor

*Published with the assistance of the
Hagley Museum and Library*

James P. Kraft

Stage to Studio

Musicians and the Sound Revolution, 1890–1950

The Johns Hopkins University Press   *Baltimore and London*

© 1996 The Johns Hopkins University Press
All rights reserved. Published 1996
Printed in the United States of America on acid-free paper
05 04 03 02 01 00 99 98 97 96   5 4 3 2 1

The Johns Hopkins University Press
2715 North Charles Street
Baltimore, Maryland 21218-4319
The Johns Hopkins Press Ltd., London

Library of Congress Cataloging-in-Publication Data will be found at
the end of this book.
A catalog record for this book is available from the British Library.

ISBN 0-8018-5089-4

To my mother and father,
Marion and Hubert Kraft

Contents

Acknowledgments

THIS STUDY has benefited from the knowledge and advice of many people. Edwin Perkins guided it through the dissertation stage at the University of Southern California with intelligence and care. Steven Ross, who also served on my dissertation committee, helped improve the early drafts in his own thoughtful ways. Richard Jewell of the USC Cinema School, Terry Seip, and Frank Mitchell also read and commented on all or parts of the dissertation drafts, and Doyce Nunis, Jr., pointed me to valuable resources. I am deeply indebted to these and other scholars at USC for their generosity and counsel.

Turning the dissertation into a book entailed new obligations. At Ohio State University I encountered an especially sharp group of business and labor historians who offered new direction and inspiration. David Sicilia read an earlier version of the manuscript and made many valuable suggestions. Mansel Blackford and Warren Van Tine went over selected chapters and alerted me to inconsistencies as well as additional sources. At the University of Hawaii I have had the opportunity to work in an equally supportive environment. I am particularly grateful to Robert Locke and Idus Newby for reading and rereading drafts of the book to my advantage.

I wish to thank librarians and archivists at the University of Southern California; the Academy of Motion Pictures, Arts, and Sciences; the University of California at Los Angeles; the University of California at Berkeley; the University of Puget Sound; the Ohio State University; the Ohio Historical Society; the Cleveland Public Library; the AFL-CIO Economic Research Department; and the University of Hawaii Library. I also wish to thank Serena Williams, Bill Peterson, Ed Ward, Sue-Ellen Hershman, Thelma Stewart, and Mark Massagli of the American Federation of Musicians; and Lee Petrillo, Milton Gabler, John TeGroen, Phil Fischer, Lenny Atkins, Eudice Shapiro, Henry Gruen, Al Hendrickson, Art Smith, Helen Lee, Jack Bunch, Roc Hilman, Richard Taesch, Rocky Jackson, Robb Navrides, Bob Fleming, Will Brady, Buddy Peterson, Bill Hitchcock, and Robert Alder, each of whom I interviewed in the course of my research.

For the photographs included herein, I am grateful to Dace Miller, Ned Comstock, Peter Lloyd Jones, Bill Benedict, Sarah Partridge, Marc Wannamaker, Greg Geary, Terry Geesken, Elvis Brathwaite, Vernon Will, Charles Arp, Elizabeth McCullough, Miles Kreugger, Carol Merrill-Mirsky, Jerry Anker, Lowell Angell, and Ray Avery. For permission to reproduce photographs, I wish to thank the University of Southern California, the Theatre Historical Society of America, the Case Research Lab Museum, the Bison Archives, the *Chicago Sun-Times,* the Ohio Historical Society, the Bettmann Archive, the Museum of Modern Art, AP/Wide World Photos, and the American Federation of Musicians. For help with tables and graphs, I thank Chris Grandy, David Ross, Billie Ikeda, Gwen Agina, and Susan Abe of the University of Hawaii, and George Potamianos and John Ahouse of USC.

The book and I have also benefited from the advice of editors and anonymous referees. Chapter 1 draws on "Artists as Workers: Musicians and Trade Unionism in America, 1880–1917," *The Musical Quarterly* 79 (Fall 1995): 512–54. Chapter 2 is a reworked version of "The 'Pit' Musicians: Mechanization in the Movie Theaters, 1926–1934," *Labor History* 35 (Winter 1994): 66–89. Chapter 4 contains material from "Musicians in Hollywood: Work and Technological Change in Entertainment Industries, 1926–1940," *Technology and Culture* 35 (April 1994): 289–314. Many thanks to Thomas Levin, Daniel Leab, and Robert Post, editors at these journals, for their guidance. I also wish to express my sincere thanks to the editorial team of Robert Brugger, Philip Scranton, and Mary Yates, who helped me turn the manuscript into a book with consummate skill.

In the course of completing this project, I received financial assistance from the John and Dora Haynes Foundation, the University of Southern California, the Ohio State University, and the University of Hawaii. On a more personal level, I wish to thank Louise McReynolds, Kenneth Lipartito, Michael Speidel, Arthur Verge, and Sheldon Anderson for their help, and all the friends who offered good cheer and moral support during the past decade. My most heartfelt thanks belong to my wife, Reneé, whose special contribution to my work is more important than words can say.

All errors of fact or interpretation, of course, are mine alone.

Acknowledgments

Stage to Studio

Introduction

THE STORY OF the harnessing of sound waves by entertainment industries is less a tale of glamour and personalities than one of new technologies, business enterprise, and workers riding roller coasters of boom and bust. Modern sound technology originated in early developments in telegraphy. In 1877 Thomas A. Edison made the first phonographic recording when he recited a nursery rhyme into a telephone diaphragm fitted with a needle that carved grooves onto a cylinder covered with tinfoil. Edison and others soon made recordings of higher fidelity using spring-driven motors, jeweled needles, flat discs, and other technological contrivances. The recording process reached new heights in 1915, when Edwin S. Pridham and Peter L. Jensen connected a power transformer and a twelve-volt battery to existing electrical circuitry, thereby dramatically increasing the volume of recorded sound.

This development—the loudspeaker—ensured the success of phonography and thus of radio. Radio enthusiasts had been sending broken messages in Morse code without telephone lines since the turn of the century, when inventors first captured the power of electromagnetic waves. The resulting "wireless" primarily served maritime interests until 1906, when Reginald Fessenden demonstrated that more powerful alternators could make "continuous wave" transmissions. By the early 1920s broadcasters were running telephone wires from radio stations to football stadiums to provide new sources of entertainment, and telephone lines were carrying programs from one station to another, and thus to audiences far removed from the original broadcast. As a result, broadcasters in small and remote communities began hooking up to powerful stations in large cities to gain access to news and entertainment programming.

As radio networks crisscrossed the nation, parallel advances in sound technology revolutionized the motion-picture industry. Inventors first tried to mesh the phonograph and the camera in 1894, when William K. L. Dickson introduced a coin-operated Kinetoscope on Broadway. Problems of synchronizing sounds and photographs, however, delayed the marriage of technologies until 1923, when Lee de Forest, drawing on the work of Theodore Case, copied recorded music onto a narrow filmstrip. By 1926 Western Electric and Warner Bros. had coupled film technology with high-quality amplifier tubes and slow-turning phonographs to produce sound movies. The instant popularity of these movies persuaded industry leaders to abandon silent films in favor of "talkies." Over the next two decades these developments and others that built onto and sometimes superseded them contributed to the rise of television, frequency-modulation (FM) broadcasting, and other forms of new and improved mass communication and entertainment.

Stage to Studio describes and assesses the impact of this sound revolution on one large but atypical group of American workers—professional musicians—during a particularly stressful time of economic and social upheaval, the second quarter of the twentieth century. The deployment of new sound technologies into the mainstream of commercial activity transformed the musicians' world, turning a diffused, labor-intensive, artisanal structure into a centralized, capital-intensive, highly mechanized one. Technological change affected wages, working conditions, patterns of hiring, definitions of skills, and above all job opportunities. It brought higher incomes and improved standards of living to many, and fortune and fame to a few; but for the majority the change meant dislocation, restricted or lost opportunity, and sustained conflict with management.

Disaffected musicians did not stand passively by while the revolution capsized their lives. On the contrary, in myriad and clever ways, and largely through their union, they sought to control the forces of change. In the decade following the introduction of sound movies, change was so rapid and overpowering that instrumentalists, in the words of one of their union officials, "did not know how to cope with this gigantic problem."[1] But once a sense of stability settled over the "music sector" of the economy, the union resisted the direction of industrial development or, more specifically, management control of new production technologies. In the 1940s, under the leadership of James C. Petrillo, musicians won major concessions from industry and in the process pioneered new patterns in labor relations. By midcentury, however, their campaign to "keep music alive" had suffered major setbacks, and they and their union were in retreat.

At the heart of this study, then, are two perennial concerns of historians of labor and technology during and after the Industrial Revolution: What impact did technological change—especially change that increased worker efficiency and productivity and thus benefited employers and consumers—have on workers? And how successfully did workers cope with that impact?

Not surprisingly, definitive answers to these questions are elusive. In one industry after another new methods of production revamped labor processes and capsized the traditional "world of workers." In many industries labor-saving machinery simplified work tasks and thereby reduced skill levels to the detriment of workers, while in others it generated demands for new skills and talents and increased the challenge of work as well as labor's bargaining power. In still other industries mechanization created new and highly skilled jobs that paid exceedingly well but fragmented the new craftworkers in ways that undermined labor solidarity and thus union effectiveness.[2] In all of these industries workers struggled, with uneven success, to control the pace of change in the workplace in order to preserve as many of their traditions, privileges, and jobs as possible.[3] The experience of musicians speaks pointedly to all of these scenarios, especially those that illustrate the ambiguous and ironic nature of the changes that technological innovation has so often produced. It suggests too that workers and their unions generally accepted innovation as inevitable, even as they tried to channel its impact to their own advantage.

The musicians' experience also illuminates the crucial role of government in shaping the impact of technological innovation on industrial development. It thus speaks to larger issues in American history: What is the actual as well as the proper relationship among government, business, and labor, especially as that relationship affects basic matters of social change? Should, can, or must the state be relatively neutral in matters affecting business, labor, and consumers, or should it intervene in those matters in behalf of one or another of the interested parties? And if it should intervene, when and in whose behalf should it do so?[4] The role of the state in industrial relations has never been static. In the 1940s business forged a closer relationship with the federal government than had been the case during the 1930s; one result of the shift was that the fate of working musicians became closely tied to politics. In this instance, at least, government policy evolved in ways that eroded labor's bargaining power generally and the ability of musicians specifically to control the impact of new technology on their employment.

The story of the resulting struggle of musicians against technological displacement is largely unknown. Labor historians have not ignored musicians, nor have they ignored the impact of technological transformation in the workplace. But they have neglected, dramatically so, the impact of technological change on musicians in their distinctive workplaces—movie and legitimate theaters, supper and dance clubs, radio stations, entertainment pavilions, and the like. Similarly, social and cultural historians have traced the emergence of mass culture in modern America, but their works invariably overlook musicians as workers in the new realm of leisure. Historians of business and technology have only begun to investigate the leisure business and have ignored altogether the conditions of its workforce. The experience of the vast majority of musicians remains distorted in romanticized accounts of popular bands, bandleaders, and singers in the glamorous and too easily glamorized early years of radio, recording, and Hollywood.[5]

This distortion is understandable. Most of us think of musicians as artists who "play" rather than work. The distinctiveness of musical labor obscures the fact that musicians work for a living and have a role in the nation's economy larger than their numbers suggest. The prominence of stars further complicates the story of the rank and file, fostering misconceptions about employment trends, especially the impact of broadcasting and recording on working musicians. It is similarly difficult to study the workplaces of musical workers, which between the 1890s and the 1950s varied too widely to encourage confident generalization. Then, too, musicians as workers had no meaningful apprenticeship and no standard for evaluating skills other than what the public would pay to see. In addition, their work was far more intermittent than that of most workers, often, even regularly, restricted to a season of no more than several months. Finally, the lines between labor and management among musicians were not always clear; indeed, instrumentalists often worked for each other.[6]

For all of these reasons the study of musicians as laboring people is necessarily interdisciplinary, drawing upon the varying perspectives and insights of business, social, and economic history as well as the histories of technology and politics. Yet no history is all-encompassing. This history of musicians as laborers is largely unconcerned with the impact of technology

4

or social change on the content or form of popular music. It does suggest, however, that changes in popular musical styles coincided with and are related to technological and institutional changes in entertainment industries. In addition, this study is not concerned with all musicians, but with working musicians in mass-entertainment industries under the capitalist mode of production. Its object is instrumentalists who earned most of their income from performances in places of private enterprise with vested interests in utilizing sound technology to maximize profits, reduce production costs, or control labor. This was the largest and most significant group of musicians in the country, but not an all-inclusive category. The study thus ignores the thousands of part-time musicians who typically supplemented their income from other sources with occasional musical performances. It also excludes musicians in symphony orchestras and other groups whose operating costs, including wages, were funded by taxation, endowments, or public donations.

The book is also interested in the development of worker institutions among musicians. Unlike workers in most mass-production industries who confronted the sudden introduction of labor-saving machinery, musicians faced the threat of mechanization *after* they had built a strong national union. Throughout the years covered by this study, that union—the American Federation of Musicians (AFM)—represented the collective voice and power of working musicians, and the union's response to technological change is thus an important part of the story told here. As large entertainment corporations used new technologies to effect greater efficiency through economies of scale, the role of institutions in the lives of musicians grew larger; not only unions but corporations, courts, and government agencies increasingly influenced their work and their well-being. The tale of musicians and sound technology is thus a story of institutions as well as of individuals and groups of people.

The story speaks to the expansive and paradoxical nature of capitalism. This dynamic mode of production, a driving force in history for more than half a millennium, has produced remarkable economic growth and innumerable examples of success. But it has also brought new and unexpected forms of uncertainty and catastrophe. In the view of Robert Heilbroner, capitalist development has been a "two-sided affair," its very dynamism having "a built-in insecurity, a self-endangering changefulness."[7] The experience of musicians testifies to the truth of this observation. It shows that even the most celebrated accomplishments of the capitalist market system can be, and usually are, accompanied by social dislocation.

Although a materialist perspective shapes this study, "nonmaterial" things also affected the lives of working musicians. The values and outlooks of musicians and their employers were so different that they precluded a mutually beneficial compromise of differences over the issues created by technological change. By the 1940s musicians and their employers were contesting more than material interests. On both conscious and subconscious levels they were competing for the moral high ground, and with it the authority to shape public perceptions of their contest. They fought their battles with rhetoric and symbols as well as with shows of economic or organizational muscle, and they did so in the press and the courtroom as well as in union halls and corporate boardrooms.[8]

To the extent that the musicians' experience is representative, it bodes ill for workers in our own age of rapid technological and institutional innovation. It suggests that the benefits of new technology will be distributed unevenly, and more or less according to power relationships between the major groups affected by technology. The story thus ends on a cautionary note. Is technological change liberating? The only realistic answers to that question would seem to be both yes and no, and it depends. Experience varies and will no doubt continue to vary. But the experience of musicians between the 1890s and the 1950s certainly challenges the uncritical assumption that advancing technology means social and material advancement, or more satisfying work.

One

Working Scales in Industrial America

Rapid industrialization in the late nineteenth century revolutionized the way Americans spent their leisure time. By separating work and play, concentrating populations in urban settings, and raising real wages for millions of workers, industrialization transformed traditional ways of leisure and recreation. Americans abandoned old customs and patterns of socializing as entrepreneurs applied new technologies to leisure-time pursuits. In the Gilded Age, entrepreneurs built lavish hotels, gaudy cabarets, and intricately adorned amusement parks to lure leisure consumers. Mocking Victorian values of thrift and sobriety, these ostentatious structures distracted Americans from the realities of industrial life.

The expanding leisure market meant unparalleled opportunities for musicians. In this era before recorded music, theater owners routinely hired orchestras, sometimes for seven days a week, to perform concerts or enliven vaudeville acts and burlesque shows. Similarly, places of entertainment from skating rinks and dance halls to hotel restaurants and fashionable watering holes featured live music on a regular basis. In addition, many instrumentalists traveled around the country with circuses, minstrel companies, and concert bands or, alternatively, found work close to home. These multiplying opportunities for musicians contrasted notably with the fortunes of skilled artisans in many other trades and professions. Throughout this era the rise of giant corporations, strong employer associations, and new labor-saving machinery undermined the status and power of millions of skilled laborers in the workplace. Like carpenters in the building trades of the late nineteenth century, musicians benefited from working for small businesses whose successes or failures depended on the performances of their employees. The threat to musicians was not mechanized factories in faraway places but their own reluctance to recognize and act on their common concerns as workers in an industrializing America. Only slowly did they come to recognize their common problems, but once they did so, they built a union strong enough to protect their rights and interests through a generation of change.

LIKE OTHER EXPRESSIONS of American culture in the late nineteenth century, music mirrored the changing times. Musical styles broke with tradition and became more complex in melody, harmony, tonality, and form. The ragtime songs of Scott Joplin, with their syncopated beats and alternating octaves and chords, like the orchestrations of John Philip Sousa, the March King, contrasted sharply with the folk, church, and classical music that embodied older but persisting musical tastes. Blues singers who slurred melodic tones no less than jazz musicians with their unorthodox chords and other improvisations reflected this upheaval in musical culture. New styles of music, with greater emphasis on rhythm, freedom, and energy, characterized the new world of cultural innovation.

As a group, instrumentalists in the late nineteenth century were as varied and discordant as the music they played. The orchestras they formed transcended age, gender, race, and ethnic differences, and skill levels as well. Many bands that played in city parks and town parades on weekends consisted entirely of amateur musicians who made their living as carpenters, clerks, or upholsterers. Fraternal lodges, schools, churches, and even extended families had orchestras of their own whose members played only "for the experience." Many businesses, including cigar, typewriter, and watch manufacturers, organized bands from among their own employees. Advertisements for musicians in local newspapers in those years testify to the high demand for workers with musical skills. In Merrill, Wisconsin, cigarmakers advertised for a "cornet man, cigar maker by trade."[1] Such advertisements were an essential feature in organizing and sustaining bands and were not unlike those by which baseball teams at the time found and held on to new players.

Many employers thought of music in the workplace as therapeutic; it soothed workers' nerves, they believed, and thus spurred production. Industrialization, as it sped up and simplified work tasks, sparked tension and thus the possibility of unrest among workers, and more and more industrialists came to see a relationship between music in the workplace and worker discipline. Such industrialists encouraged employees to organize musical groups to perform at lunch breaks, work stoppages, or other times of the day. In the early twentieth century some industrialists even established music departments to help workers form bands and acquire instruments. These developments were part of a growing effort to minimize labor conflicts by improving worker morale. Company-sponsored sports teams, pension plans, and English-language classes were other manifestations of a kind of "welfare capitalism" that eased industrial relations without affecting relations of power.[2]

The proliferation of musical groups in the workplace further blurred the distinction between amateur and professional musicians. In the late nineteenth century many "amateurs" played for wages, and "professionals" often held nonmusical jobs. But the social profile of those who made their living in music and looked upon it as a livelihood rather than a hobby differed considerably from that of amateurs. Professionals had higher skill levels, usually attained through private lessons, institutional training, work experience—or sheer talent. Most also had higher career ambitions, greater commitment to music as an artistic as well as a professional enterprise, and a better sense of music as a business. These traits were of course necessary in a trade in which workers frequently changed employers and just as frequently faced prolonged periods of unemployment. The hardships of travel as well as the poor working conditions under which musicians often labored undoubtedly kept some instrumentalists from pursuing professional careers.

The experiences of Nels Hokanson, a seventeen-year-old trombonist who worked in Bosco's Traveling Circus at the turn of the century, illustrate why some instrumentalists shied away from musical careers. In the days before automobiles Hokanson traveled by horseback from town to town, often encountering bad weather and poor accommodations along the way. The young musician played two shows six days a week and in each new town marched in a public parade to advertise the circus. Between performances he helped set up the tents and distribute publicity notices. Such a schedule discouraged all but the most dedicated—or needy— instrumentalists.[3] The working conditions of those who stayed closer to home were not necessarily better. In Chicago, for example, where annual earnings of professional musicians equaled those of other skilled laborers, the conditions of work were often far from ideal. Musicians in vaudeville

Musicians traveling with a circus during the 1890s pose in front of their ornate band-
wagon. *(Bettmann Archive)*

theaters had arduous schedules, performing as much as two matinées and
a nightly show six or seven days a week. "Fiddling or drumming or sawing
a big brass [instrument] may not look like hard work when viewed from
the comfortable balcony chair," one Chicago musician said, "but it is hard
work, monotonous as well, and exacting." Those who worked in dance
halls, hotels, and ballrooms also complained about their working environment. "You will usually find the orchestra," a Chicago trade paper complained of such places, "stuck up in some gallery or loft, ill-ventilated,
where [musicians] are continually taking into their lungs the heated viti-
ated air which creates thirst, and from this musicians are accredited with
being a drinking set."[4]

Yet poor working conditions affected musicians much less than they affected other skilled workers. The sheer love of performing and the desire
to improve their skills led musicians to make full-time commitments
whenever possible, regardless of the conditions under which they worked.
As job opportunities rose, this dynamic helped boost the ranks of full-time
instrumentalists. Census takers recorded a dramatic rise in the number of
musicians performing for a livelihood. In 1870, when the nation's popula-

tion was under forty million, sixteen thousand men and women listed their occupation as "professional musician" or "teacher of music." In the ensuing decade, while the nation's population increased 30 percent, the number of musicians and music teachers doubled. By 1890 the figure had doubled again, to over sixty-two thousand, and it reached ninety-two thousand in 1900.[5]

ONE YARDSTICK OF the resulting increase in professionalization was the extent of unionization. Worker cooperation for purposes of protecting wages and working conditions can be traced back to colonial times, when "benevolent and protective" associations of printers, carpenters, and other craftsmen endeavored to regulate prices and apprenticeship programs. With the rapid triumph of the market in the Jacksonian era and the subsequent organization of industry on a national scale, workers formed the first modern trade unions to bargain with employers and protect employment conditions. In 1865 about two hundred thousand workers, 2 percent of the nonfarm workforce, belonged to such trade unions, whose leaders generally accepted the capitalist order but were prepared to act collectively in behalf of what they considered the vital interests of their members.

Unionization among musicians dated back to the late 1850s, when musicians in Baltimore and Chicago formed fraternal organizations for self-protection as well as mutual assistance. Similarly, musicians in New York City in 1860 established the Aschenbroedel Club to promote "the cultivation of the art of music . . . good feeling and friendly intercourse among the members . . . and the relief of such of their members as shall be unfortunate."[6] The New York club assumed something of the nature of a trade union in 1863, when it changed its name to the Musical Mutual Protective Union (MMPU) and received a state charter limiting its legal liability as a public corporation. Following this example, musicians in Philadelphia organized their own musical association in 1863, and those in Washington, St. Louis, Boston, Cincinnati, and Milwaukee soon did likewise.[7]

These early organizations were more labor exchanges or hiring halls than unions per se. Like inside contractors in factories at the time, musicians met in their own "union" halls to buy and sell their services. Through the agency of the union, buyers, known as leaders, purchased the services of sidemen—musicians who accompanied a leader in a band, ensemble, or whatever. A musician became a leader by contracting the proprietor or manager of an entertainment facility for the employment of a group of musicians. Given the informality of such arrangements, it was not unusual

Table 1 List of Prices for Musical Services, Cleveland, Ohio, 1864

1. To play in front of a hall for one hour $1.50
2. Saloon Concerts $2.00
3. Serenades for one hour $2.00, each subsequent hour $1.00
4. Funerals to Erie Street Cemetery $2.50, to Woodland or West Side Cemeteries $3.50
5. Escort of any corpse to or from different depots $3.00
6. Parades $3.00
7. Banquets $4.00
8. Weddings $4.00
9. Private Parties $4.00
10. Moonlight Excursions $4.00
11. Political Meetings for one hour $2.00, whole evening $3.00
12. Picnics on weekdays—all day $5.00, half a day $3.00
13. Political Excursions for one day $6.00, for several days $5.00
14. Balls $5.00, on holidays $6.00
15. German Kraenzehen till 1 o'clock $3.00; each subsequent hour $.50
16. Fairs with dance till 2 o'clock $4.00, if longer $5.00
17. Public Concerts with one rehearsal $4.00, extra rehearsal $1.00
18. Masses with one rehearsal $3.00, extra rehearsal $1.00
19. Dancing till 12 o'clock for Scholars $3.00, till 2 o'clock $4.00
20. Opera per Evening $3.50, per week $18.00

Source: "The Musical Association, 1887–1912," pamphlet in Bagley Collection.

Note: These prices, established by the Cleveland musicians' union in 1864 and printed in the local's handbook for members, reveal the unique and varied nature of the musicians' work environment and show how local unions carefully controlled wages and the length of performances. Prices are for the services of one musician.

for individuals to shift back and forth from leader to sideman. In addition to supplying the mechanism for bringing leaders and sidemen together, these early unions acted as fraternal support systems, providing members with modest pensions, credit arrangements, health insurance, and death benefits.[8]

The most important objective of the unions in the late nineteenth century was to establish and maintain uniform wage scales. To establish the desired uniformity, local unions created price lists specifying acceptable wages for various types of work. The prices depended on the kind as well as the duration of the work involved. For steady employment in circuses, vaudeville, cabarets, and the like, the unions generally prescribed weekly wages. For one-time occasions such as weddings and private parties, fees were lump-sum. In 1865, to illustrate, members of the Washington [D.C.] Musical Protective Union charged concert saloons $30 a week for leaders

Stage to Studio

and $16 for sidemen, while the charge for a single masquerade ball or town parade was $6 per musician. A generation later, in 1891, the Columbus [Ohio] Musicians Protective Association prescribed wages of at least $2 per musician per engagement at skating rinks, baseball games, and funerals, with an additional dollar for bandleaders.[9]

Like other wage earners, union musicians established their own customs concerning the pace and duration of work. Price lists carefully specified the hours of work. Musicians working on steamboats in Washington in 1865, for example, received $6 each for "Moonlight Excursions" that lasted from "8 until 4 o'clock." For excursions that went beyond four o'clock, they received extra pay by the hour. As in all trade unions, many workplace customs were left unstated in union rules. Periodic breaks, for example, punctuated the schedules of all musicians and allowed them as well as their audiences to converse and refresh themselves. In Columbus the union price list for string performances specified, "Refreshments to be served in all cases."[10]

Local unions organized themselves democratically. They vested executive power in elected officials, whose numbers and titles varied according to the size of locals but who usually included a president, a vice president, and a secretary-treasurer. Officials typically served one-year terms but were often reelected. Presidents generally appointed business agents, who recruited new members and policed the activities of members and their employers. Membership meetings, which convened monthly, quarterly, or annually, constituted the principal instrument of governance. Such meetings, however, were not typically well attended, and small groups of dedicated members often dominated union affairs. Most members attended union meetings only when special problems or opportunities arose. Between sessions executive boards exercised the powers otherwise reserved for the membership meetings. The boards also had power to discipline and expel members.[11]

Not all musicians joined a union. Some made their own terms with employers without regard to union rules or pay scales. To combat this, union officials in large cities negotiated all-union hiring agreements with proprietors, who came to terms because they needed regular access to popular musical groups whose members belonged to the union. Unions pressured professionals to join their ranks by fining members for performing with nonmembers, a practice that fostered union solidarity and closed-shop hiring contracts. As a result of such strategies, by the late nineteenth century most musicians found it necessary to join a local union. Some instrumentalists, of course, needed no pressure to do so. Many of them came from families with union backgrounds, and joining a union was for them a natural thing to do.[12]

For perhaps half of all union members in the late nineteenth century, music was not the sole source of income. Some musicians joined unions because doing so facilitated the earning of supplementary income, while others joined in order to meet other musicians in professional settings. Unions accepted marginally talented members, because ambitious, nonunion amateurs could undercut union price lists. A few of the oldest unions in large eastern cities required professional competence for membership, but most locals did not. In the latter, applicants for membership might have to play a song or answer a few questions, but most had only to pay an initiation fee and agree to abide by union rules.[13]

This disregard for professional standards diminished union credibility in the sense that it ran counter to general trends among other skilled workers. But it was a reasonable response to a unique problem: musicians' unions were in no position to certify the professional quality of their members. On the job market, sight-reading aptitude might prove less important than improvisational skills, individuality of interpretation, or even stage presence. Diverse and changing tastes in popular music also made it difficult to justify any policy of exclusion. Appraising marketable entertainment skills was a highly subjective art, and that fact among others set musicians' unions apart from professional organizations or artisans' unions, which used internally imposed standards to protect themselves in a competitive market economy.

UNIONIZED MUSICIANS generally shared the social prejudices of other turn-of-the-century Americans; these prejudices discouraged solidarity among musicians. Despite the impressive contributions of African Americans to musical culture, to illustrate the problem, black musicians generally could not join white unions. This of course caused problems on both sides of the color line. In Cleveland, for example, white union officials complained that "unschooled" black instrumentalists controlled a large share of the local "dance and party business." Like workers in other segregated unions (or trades), white musicians in Cleveland saw their black counterparts as "industrial competitor[s]" and "treated [them] as such."[14]

To protect themselves, African Americans in large cities often organized their own musicians' unions, which maintained loose contacts with their white counterparts, but had their own union halls, officials, and arrangements with proprietors. In 1875, for example, black musicians in Boston formed the Progressive Musical Union and established price lists roughly equal to those already established by white musicians. In many cities without black unions, African Americans created musical agencies that served as union stand-ins. Many of these agencies evolved from informal contacts at music stores, dance clubs, or even poolhalls. One such agency was the Clef Club in New York City, where black instrumentalists met and contracted for each other's services.[15]

The difficulty of forming unions did not prevent African Americans from distinguishing themselves as professionals in the late nineteenth century. Some black musicians attained national and even international fame as concert musicians; others were prominent in commercial music. During the 1870s and 1880s many African Americans received musical training in army bands and made the best of it afterward in theaters and nightspots. New Orleans became famous for its unsurpassed brass bands made up entirely of black musicians. Minstrel troupes were another source of revenue for black musicians, especially banjo players and guitarists. The male-dominated black minstrelsy became one of the nation's favorite forms of entertainment, especially in the South, where minstrel companies traveling by rail made seasonal tours performing comedy, dance, and musical acts under big canvas tents. Among the popular touring groups were Brooker and Clayton's Georgia Minstrels, Silas Green, and the Rabbit Foot Minstrels. Their accomplishments are all the more impressive because they were made against the backdrop of pervasive racism in the "the age of segregation."[16]

The professional achievement of black musicians rose following the exodus of African Americans from the South that began in the 1910s and accelerated until the Great Depression of the 1930s slowed it for a time. Aspiring black instrumentalists were among the first to abandon the South, where musical jobs were few, employment uncertain, and income generally too low to sustain a musical career. In the 1870s and 1880s, before the exodus began, black musicians congregated in southern towns along railroad trunk lines, like Memphis (Tennessee) and Jackson (Mississippi) on the Illinois Central. By 1900, when nearly a quarter of all African Americans lived in urban areas, most of them still in the South, many of these musicians had moved to Chicago, Detroit, and other northern cities, where larger audiences supported their music. The musicians often worked in the saloons or vaudeville theaters in red-light districts known for gambling, prostitution, and other illicit activity. Some worked not for wages but for tips from the audience and for food and drink from the proprietor. The fact that black musicians brought their own styles and repertoires to southern and then northern urban centers left a deep imprint on popular American music. Their folk songs and blues and rag styles, and especially the unique improvisations of their jazz, became popular across America, among black and white audiences alike.[17]

The experience of Henry Thomas, a self-taught guitarist from Upshaw County, Texas, illustrates these general patterns. Thomas left home around 1890 and began playing guitar in train depots in east Texas. With rousing, foot-stomping songs like "Alabama Bound" and "Old Country Stomp," Thomas made a name for himself locally and enough money to survive. By 1893 he had traveled as far north as St. Louis, working wherever he could along the way.

Thomas was known for the unique style he created by pounding his thumb on the guitar to produce hard beats while he picked melodies high on the guitar neck and ran a knife blade along the strings to produce a distinctive twangy, bluesy sound. He also played the quills, a predecessor of the harmonica made of cane reeds cut to different lengths and tied together in a row. Thomas blew across the top of the reeds to produce his signature high-pitched melodies. The opening lyrics to one of his songs, "Railroadin' Some," speaks of his work experience:

> I leave Fort Worth, Texas, and go to Texarkana
> And double back to Fort Worth
> Come on down to Dallas,
> Change cars on the Katy
> Coming through the Territory to Kansas City,
> And Kansas City to St. Louis
> And St. Louis to Chicago
> I'm on my way but I don't know where.[18]

Women musicians faced their own set of challenges in the male-dominated world of turn-of-the-century America. Social custom discouraged female instrumentalists from performing in public or for wages, and those who did perform faced prejudices that limited their career opportunities. Bandleaders and proprietors alike considered women performers less ambitious and less reliable than men, and more likely to cancel performances and quit orchestras. Many also thought women lacked the physical stamina to travel and perform every day; those who did so, it was believed, needed special attention and were thus more trouble than male

Stage to Studio

performers. Union locals, especially those that grew out of all-male marching bands, did not always accept women either.

The exclusion of women from musicians' unions paralleled the general trends in unions at the time, especially unions of skilled laborers. Throughout the Gilded Age and Progressive Era, far fewer women than men belonged to trade unions. In 1900, for example, only 3 percent of women working in industry were union members, compared with nearly 20 percent of men. Trade unions at that time made little effort to organize women, and women who did join unions were unable to participate in union meetings and governance on an equal footing with men. Most trade unionists believed that female workers drove down wages and robbed male workers of "dignity" and "backbone." In matters of gender equity the record of musicians' unions was thus better than that of labor organiza-

California Women's Symphony, Los Angeles, 1893. Despite being relegated to the margins of the profession, all-female orchestras at the turn of the century often played at public recitals and private parties. *(Hearst Collection, Department of Special Collections, University of Southern California)*

tions in general. The constitution of the Columbus local reflected the general attitudes of male musicians: "Whenever the word 'man' occurs," the charter stated, "it shall be so construed as to mean 'woman,' 'musician,' or 'member,' and when 'he' or 'him' occurs, it shall also mean 'she' or 'her.'"[19]

Despite their marginalization in musicians' unions, women played musical instruments and otherwise contributed to musical culture as well as the business of music. Some women played for wages alongside men, while others performed alone or only with other women in public recitals, or in socially sanctioned settings such as church or at home. A few joined all-women orchestras, which occasionally performed in public. The twenty-one-piece California Women's Symphony, for example, performed regularly in the late nineteenth and early twentieth centuries. The role of women in music education was much more prominent and important than in musical performance. According to the 1890 census, about 60 percent of the nation's sixty-two thousand musicians and music teachers were women. Since men dominated the union membership lists of performance musicians at that time, that figure suggests that women were much more involved in music education than were men. This too was distinctive, if not unique, among skilled workers at the time: those who taught workers their skills were not themselves considered skilled workers.[20]

Perhaps the instrument played most frequently by women was the piano. Many women apparently believed that the piano was physically easier to play than stringed instruments and horns, and they and others seemed to agree that it was a "proper" instrument for women. Popular magazines and mass-circulation newspapers, as well as ladies' journals, told women that the piano was intimately linked to "true womanhood," that is, to notions about women's proper place in society. Pianos helped women fulfill their natural roles as ornaments of the home and the family. Playing the piano properly also solved women's "posture problem." While playing, women kept their backs straight and their knees and feet together, and thus looked "feminine" and "cultured." "There she could sit," one handbook on etiquette explained, looking "gentle and genteel, . . . an outward symbol of her family's ability to pay for education . . . of its striving for culture and the graces of life, of its pride in the fact that she did not have to work and that she did not 'run after men.'"[21]

There were still other signs of division among musicians. Language and ethnic barriers impeded worker solidarity, even cooperation, in the music business just as they did in other trades during the Gilded Age. Census reports and obituaries in union trade papers indicate that while most musicians were native-born, a very large minority—perhaps a third—were not. Substantial numbers of instrumentalists in many urban centers spoke German, while many others spoke a Scandinavian or an eastern European language.[22] In the 1880s, a time of heavy migration from eastern and southern Europe to the United States, union officials in Cleveland noted that the city's musicians were divided between "those who spoke English and those who spoke German." This statement reflects the fact that a recent influx of immigrant musicians from Bohemia had made communication difficult between instrumentalists in the city. Even Bohemian musicians, labor organizers discovered, were split into factions.[23]

This lack of solidarity affected all musicians negatively. In Cleveland, white union officials complained that competition between ethnic groups had caused wages to drop "almost out of sight." "Employers were the only ones who gained," one official explained, "in [this] merry war betwixt tweedle dee and tweedle dum." White locals encouraged European immigrants of all nationalities to join their ranks, but despite their efforts, ethnic diversity undermined union power in Cleveland and elsewhere throughout the late nineteenth century.[24]

Beyond these divisions and tensions within the workforce lay looming problems of an even more intractable nature. Throughout the Gilded Age and Progressive Era, the movement of musicians from one locality to another as well as the hiring of amateur, immigrant, and military bands meant growing competition for local professionals. In addition, periodic economic downturns disrupted the demand for musical services and encouraged musicians, like other workers, to undercut each other's wages. At the same time, new musical styles and new generations of musicians and music consumers put constant pressure on instrumentalists to retool, as it were. To attract new customers or increase the sale of food and drinks, proprietors had few qualms about dismissing established groups when more stylish, innovative, or newly popular groups became available. Such practices showed musicians the advantages of organizing nationally as well as locally.

MANY SKILLED WORKERS had already learned those advantages. Printers had formed the first national labor union as early as 1852. Twenty years later, when nearly three hundred thousand workers belonged to trade unions, no fewer than thirty artisanal groups boasted of national organizations. National organization helped local unions deal with the structural and legal problems of unionization. At the same time, the full-time offi-

Joe Sheehan's Orchestra, circa 1910–20. Dance bands such as this one played music in theaters and dance halls in the early twentieth century. *(Ohio Historical Society)*

cials who ran the national unions gained valuable experience in dealing with employers, courts, and legislators.

The first call for a national organization of musicians came in November 1870, when the Philadelphia Musical Association, concerned about the growing presence of immigrant and nonunion musicians, proposed a "general union" of local "musical protective associations." Acting on the proposal, delegates from associations in New York, Boston, Baltimore, Chicago, and Philadelphia met in June 1871 and established the Musicians' National Protective Association (MNPA). Essentially a loose confederation of independent locals, the MNPA lacked both the power and the resources to accomplish much. Despite the enthusiasm of its leaders and the apparent strength of some of its affiliates, the MNPA was unable to sustain the momentum that had given it birth. In the depression of the 1870s it disintegrated, as did twenty other national labor organizations.[25]

When the economy rebounded at the end of the decade, so too did national labor organization. A second, more successful union of musicians emerged in 1886, when the members of a Cincinnati local contacted locals in other cities and proposed the formation of a national union. In March of that year delegates from musical societies and union locals in New York, Boston, Philadelphia, Chicago, Detroit, Cincinnati, and Milwaukee met at the Grand Hotel on Broadway in New York City and established the National League of Musicians (NLM). The league was another federation of local groups designed to promote fraternal relations as well as to protect wage levels and improve working conditions. Its first president, Charles M. Currier, was an accomplished musician who had once worked in the well-known Gilmore Band in Boston and in other groups from New Orleans to Chicago and Cincinnati. Currier and others in the league hoped that a national organization would persuade Congress to pass legislation prohibiting musical proprietors from hiring foreign orchestras or military bands for performances for which local groups were available. They argued that proprietors hired such groups to avoid paying union prices, and they promised to expose "the theatrical landsharks and managerial swordfishes" who exploited musicians.[26]

For a while the NLM grew rapidly. Ten years after its founding, it had more than one hundred affiliates, but its strategies as well as its structure were ill suited for the changing times. The depression of the 1890s, competition from foreign and military orchestras, and growing numbers of nonunion musicians put downward pressure on wages and compromised NLM price lists. Local unions could not remedy the situation, and the NLM had no independent power of its own. In fact, its structure as well as its purposes generated dissent, especially among smaller and newer locals in midwestern states. By an unusual proxy system, delegates to its annual conventions from large and well-established locals in the East voted on behalf of locals too small or poor to send delegates of their own. In this way the New York local, whose leaders were mostly insensitive to the needs of working musicians in small cities, came to dominate the conventions.[27]

The attitude of prominent NLM officials toward trade unionism in general was an even greater problem. Since the 1860s organizations of skilled workers had tried to forge a national association of unions to represent the common interests of unions and workers. They finally did so in 1886, when a convention of trade union representatives in Columbus founded the American Federation of Labor (AFL). This umbrella federation had little power relative to its strongest national affiliates, yet it offered them distinct advantages. AFL leaders promised to promote organizational drives, mediate jurisdictional disputes between rival unions, and keep track of legislation and legislators of interest to member unions. Unlike its predecessors, the AFL embraced the basic premises of capitalism; its goal was to help skilled workers secure larger shares of the material rewards of capitalism than they had secured in the past. The federation therefore focused its attention on the issues of wages, hours, and working conditions rather than on larger matters of social reform. Federation president Samuel Gompers wanted affiliates to attain their goals by collective bargaining, but he was ready, when necessary, to support strikes with whatever resources he had available. Gompers hoped to make the AFL the most inclusive confederation of unions in the nation and accordingly met with leaders of all national unions, including the musicians.

In 1887 Gompers addressed the NLM convention, urging musicians to join the trade union movement. Despite the evident interest of musicians in doing so, league leaders rejected the offer. When Gompers reiterated the invitation in subsequent years, they continued to rebuff it. NLM officials from large East Coast locals were no doubt worried about losing their own power and privileges as leading spokesmen for musicians, but they also believed that musicians were artists and not workers and had little in common with members of the AFL. In any case, they argued that affiliation might cause musicians-as-artists to suffer a loss of dignity.[28]

The league's response must be viewed in the context of the times. The late nineteenth century was a chaotic, truculent era in industrial relations, a time of sweeping innovation, unrestrained competition, and unbridled industrial growth. It was also a time of reactive labor militancy and class violence. In industry after industry, new methods of production revolutionized the labor process and capsized the traditional world of workers. When labor responded to management heavy-handedness with strikes or other forms of protest, management countered with lockouts, strikebreakers, and even organized violence. Many strikes ended with the intervention of local police, state militia, or even, occasionally, federal troops. In 1887, the year Gompers invited the league to affiliate with the AFL, there were, according to imperfect figures, 1,436 strikes in America involving 273,000 workers. Over the next ten years, according to the same imperfect sources, there were another thirteen thousand strikes involving 2.5 million workers. From the steel mills of Homestead, Pennsylvania, to the silver mines of Coeur d'Alene, Idaho, organized and unorganized workers resisted the new industrial order, only to discover in the process that local, state, and even federal authority protected it. This turbulent world of dynamic corporations, labor-saving machinery, displacement of skilled workers, and industrial conflict was largely removed from the experience of musicians before the turn of the century, and that fact nurtured the conviction among them that they had little in common with other workers as workers.

But the NLM response to Gompers's invitation reflected widening divisions among musicians, growing concerns about group identity, and even emergent class consciousness. Musicians in societies in several large East Coast cities had traditionally thought of themselves as an elite, a highly creative group of artists. They distanced themselves not only from industrial workers but from other musicians as well. Those among this elite who performed works of classical composers in symphony halls and well-to-do neighborhoods not only had years of formal training as well as highly developed skills, but like their audiences, they believed that classical and even semiclassical music was better, more culturally enriching than popular music performed in less sanitized places. In other words, a self-validated elite proclaimed and perpetuated a hierarchy in musical culture that had implications for the labor history of musicians, sustaining as it did a division between "refined" music and musicians on the one hand and folk, country, and black music and musicians on the other. Unlike cigarmakers and machinists, elite and "respectable" musicians reasoned, musicians did not labor; rather, they performed. True, musicians already had local unions and work rules, but their lives seemed, to most musicians as to most economists, largely removed from the world of labor economics. Collective bargaining no less than collective activity like strikes seemed antithetical to matters of musical performances, to say nothing of musical art and beauty.

A new generation of musicians in the 1880s and 1890s began to feel differently about these matters. In contrast to the established elite, many young musicians, especially those in popular or mass forms of music, had working-class backgrounds and depended on wages and on the business of music to earn a living. If they recognized a musical hierarchy, as they probably did, they did not therefore discredit, or even discount, the value of their own activity as musicians. They preferred the music they performed, as did their audiences. As that suggests, art and aesthetics were to them things mediated by popularity and market demand. They wanted to capitalize on those things and saw in the organization and tactics of skilled trade unionists models of self-protection and advancement they might use to advantage.[29]

Responding to these developments, Owen Miller, the president of the NLM, told musicians in 1895 that they must reorient themselves and the league toward trade unionism. Only in that way, Miller insisted, could the league and its members respond meaningfully to the challenges that threatened their economic well-being.[30] Miller's call for radical change, however, fell mostly on deaf ears. A year later a new president, Alexander Bremer, who was also head of the New York City local, sounded a much different note. Seeking alternately to obscure and to trivialize Miller's concerns, Bremer insisted that musicians would not "cast their lot" with "stovemolders" and "shoemakers." The differences exemplified by the conflicting positions of Miller and Bremer were fought out in the councils of the league. Bremer proposed in 1896 that league conventions meet biennially rather than annually, and that between conventions the executive board be empowered to exercise "a general supervision of all matters pertaining to the League." This would of course entrench the eastern elite at the expense of the younger dissidents, who were already convinced that Bremer and his supporters wanted to mute their voices and ignore their protests. The dissidents concluded that league leaders—whom they derisively called "Silk Hats" and "Prince Alberts" because of the fancy clothes they wore—would never change.[31]

One immediate consequence of this division was strengthened ties between some league locals and the American Federation of Labor. In 1895 Gompers announced his willingness to support a new union of musicians if the NLM refused to affiliate with the AFL. At the same time, he renewed his offer to make a reformed league an autonomous affiliate of his federation. The league convention rejected this offer by a tie vote, whereupon Gompers scheduled his own convention of musicians to meet in Indianapolis in October 1896.[32] A total of twenty-six locals, including seventeen affiliates of the NLM, sent delegates to Gompers's convention. There, C. H. Ruhe, chairman of the league's executive board and one of the delegates, persuaded some of the delegates to withdraw from the convention. But those who remained represented approximately three thousand musicians, and out of their deliberations came the American Federation of Musicians (AFM). The structure of the AFM was like that of other affiliates of the AFL. National and local organizations shared powers. At the national level there were nine elected officers—president, vice president, secretary, treasurer, and five executive committee members—who oversaw the affairs of the union. As was generally the case in unions, the president had considerable power, including the power temporarily to "annul and set aside" provisions of the union constitution except those dealing with finances. Ultimate authority in the AFM as in other AFL unions, however, rested in the annual conventions, at which delegates from locals set national policy and elected national officers for terms of one year. Each local had one vote at the conventions for every one hundred members, but to limit the power of large locals, no one of them had more than ten votes. The conventions and national officers had joint responsibilities in resolving disputes within the federation, for which purpose the executive board acted as an appellate court.[33]

The first president of the AFM was Owen Miller, who had earlier sought to bring the NLM into the AFL and, when that failed, had taken the lead in forming the AFM. The choice of Miller, who was forty-five years old in 1896, was thus appropriate. He was a widely respected figure in politics and labor, having once served in the Missouri state senate and as a leader of the Missouri Federation of Labor. His lack of formal education he compensated for by breadth of experience and astuteness of judgment.[34]

Miller immediately found himself and his new union in a power struggle with the NLM. Embracing the goals and methods of other AFL unions through a relatively democratic organizational framework, Miller sought to make his new union the organized voice of working musicians. Representing instrumentalists who were more likely to work in dance halls, saloons, or other places of popular entertainment than in ballrooms or symphony halls, and who belonged disproportionately to small locals outside the urban Northeast, Miller and other AFM leaders nevertheless saw the immediate need to establish the union in large eastern cities, where employment opportunities as well as sources of union power were disproportionately concentrated. Placed on the defensive by the appearance of the AFM, leaders of the NLM sought to parry the union's appeal by appeals of their own to musicians as artists.[35]

The choice of Kansas City as the site of the AFM's second convention, in 1897, was shrewd, for the NLM had announced its intention to hold its own convention at the same time and in the same hotel as the AFM. This had the ironic effect of putting delegates to the AFM convention in a position to disrupt the NLM convention, both of which convened on May 4. More than forty delegates to the league convention were members of the AFM, compared with perhaps twenty delegates who belonged solely to the NLM. Recognizing the threat these figures posed, league officials hastily revoked the charters of all its locals affiliated with the AFM and stationed policemen at the doors of its convention hall to make sure no one entered as a delegate without credentials signed by league officials bent on excluding delegates who also belonged to the AFM. This strategy collapsed when a group of excluded delegates obtained a court order directing league officials to admit them to the convention. The frustrated president of the league promptly adjourned its convention.[36]

Revoking the charters of locals affiliated with the AFM effectively killed the NLM. When the adjourned convention reassembled a year later only nine locals sent delegates, and that number declined to three in 1902, when delegates from New York, Philadelphia, and Baltimore were the only ones to appear. The following year the league's New York chapter merged with the city's growing AFM local, and the chapters in Baltimore and Philadelphia followed suit. What remained of the league following these actions quietly disbanded in 1904.[37]

The rise of the AFM was basically a restructuring of the national musicians' union, since many leaders of the new AFM had been prominent, if often dissenting, members of the NLM. But the rise of the AFM also signified an important change in the social outlook of working musicians. Impressed by the methods other skilled workers were using to protect or better their interests, musicians began discarding the image of themselves as artists or performers and replaced it with a growing consciousness that they were skilled laborers with interests and circumstances of their own as working people. They came to believe, in other words, that only a powerful, assertive union committed to labor solidarity and willing to challenge management when necessary could protect and advance their collective interests as musicians.

While not denying that musicians were artists and professionals, the AFM embraced the strategies and tactics of skilled trade unions in pursuing its objectives. More aggressive and confrontational than the NLM had been, the AFM was more willing and able to challenge entrepreneurs who employed musicians. Its democratic structure and policymaking processes helped the federation maintain the loyalty of its members, and its activist style and pragmatic purposes facilitated rapid growth. Absorbing most NLM affiliates at the Kansas City convention in 1897, the AFM then boasted 72 locals; the number reached 114 in 1900, when the union began absorbing Canadian locals. At the outset of the twentieth century the union had more than ten thousand dues-paying members and was the undisputed voice of working musicians.[38]

The artist-versus-worker controversy persisted, however. Indeed, the idea that musicians had little in common with other workers remained strong and functioned to limit the union's strength throughout the period of this study. But by 1900 the controversy no longer prevented the building of a national union that embraced the philosophy of the skilled labor movement represented by the AFL.

The rise of the AFM cannot be understood simply in terms of pragmatics, however. Like worker organizations in the building trades, the

AFM succeeded chiefly because it confronted literally thousands of small, unorganized employers who had neither the resources nor the know-how to unite and resist union demands. Most trade unions at the time, even those of skilled workers, were less fortunate, for they emerged—when they succeeded at all—in the face of monopolistic or relatively unified employers with formidable power in the marketplace and influence in local if not national political councils. The economic power of unified musicians was relatively strong, in contrast, and their unionization campaigns met little effective opposition. The AFM took advantage of this circumstance to tighten its grip on musical services by imposing price lists and regulating working conditions. By the end of the first decade of the new century, the ability of the AFM to deal effectively with employers was as impressive as that of any union in the AFL.[39]

In 1900 Miller stepped down as president of the AFM to become its secretary as well as editor of its trade journal, the *International Musician*. The fact that the secretary's salary was $750 a year and the president's only $100 apparently caused the first of these switches. Though Miller was president of the union for only four years, his tenure was significant. The forceful, self-taught leader from Missouri played an important role in solidifying the union's position among musicians, just as he had in eliminating the NLM.[40]

On Miller's recommendation, delegates to the 1900 annual convention elected Joseph N. Weber president of the union, a position Weber held for forty years. Born in 1863 in the village of Temesvár in present-day Hungary, Weber migrated with his parents to New York City as a boy. He learned music from his father and as a young man made his living as a clarinetist in touring bands that took him to Chicago, Kansas City, New Orleans, and elsewhere. In 1891 he married a violinist, Gisela Liebholdt, and together they traveled and worked in cities on the West Coast from Los Angeles to Seattle.[41] Apparently Weber's first significant union activity occurred in Denver in 1890, when he helped organize what became Local 26 of the NLM. On the West Coast he served for a time as vice president of the Seattle local. In 1895 he moved to Cincinnati, where his father operated a saloon. There he continued his union activity, this time as president of the Musicians' Protective Union, Local 3 of the NLM, which under his leadership played a central role in organizing the AFM. Weber represented the Cincinnati local at the AFM conventions in 1899 at Milwaukee and in 1900 at Philadelphia, at the latter of which he became president of the organization.[42] His elevation was significant in part because he had no patience with those who debated the artist-versus-worker question. "We musicians are employed under the same conditions as any other workers," he told Denver musicians in the late 1890s. "We may be artists, but we still work for wages. . . . [We] are exploited by our employers in the same manner as any other wage-earners who stand alone. Therefore we must organize, cooperate and become active in the economic field like other workers."[43]

The years following Weber's accession to the presidency of the AFM were years of growing opportunity for musicians. Industrialization continued to draw people into urban centers and to increase the demand for entertainment once they were there. As a result, the demand for talented musicians often outstripped the supply, and union locals struggled not against unemployment but against what they and their members called "unfair competition" from nonunion musicians who undercut union prices and took jobs away from unionized musicians. The national union necessarily gave locals considerable freedom in handling these problems, while it acted as vigorously as it could against such nationwide problems as competition from military bands, foreign orchestras, and musical groups traveling from one local jurisdiction to another. It also did what it could to address the intractable problem of race relations.

The use of military bands in performances for which civilian musicians would otherwise have been used—and paid—was a problem of long standing. Musicians objected to the practice because military bandsmen were paid by the military and thus worked for lower wages than civilian musicians. The competition was doubly unfair because military musicians received their instruments and uniforms at taxpayers' expense.[44] How much military bands undermined the wages and employment of union musicians is difficult to determine, but in 1888, instrumentalists in New York compiled a list of over one hundred instances of military bands performing in circumstances in which civilian bands would have otherwise been employed. An instructive example of how military bands undercut civilian musicians involved the hiring of a band for the Pure Foods Show in Washington, D.C., in 1895. A civilian band offered to play at the show for the union price of $24 a week, only to be underbid by the Fourth Artillery Band, which offered to perform for $18 a week. To get the job the civilian band then agreed to play for $14 a week, but after three weeks the management of the show dismissed the band and replaced it with the artillery band, whose members agreed to perform for $8 a week, one-third of the union scale.

As this incident suggests, the low pay of service musicians encouraged them to seek outside work. AFM officials, who were fully aware of this situation, accordingly worked to raise the wages of military bandsmen. The union urged Congress and the Departments of the Army and the Navy to prohibit military musicians from competing with civilians. Perhaps in response to union efforts, in 1908 Congress increased the pay of enlisted musicians and prohibited army and navy bands from competing with the "customary employment" of their civilian counterparts. The act proved to be ineffective, however, because enforcement was left to local commanding officers, who continued to allow bands to accept outside employment. The attorney general further weakened the law by exempting the best military band of them all, the Marine Band, from the provisions of the law on the grounds that it was not an army or navy band. Not until 1934, when the navy alone had over 150 musical groups in its service, did the military, under pressure to help alleviate the nation's unemployment problems, end the practice of allowing service bands to compete with civilian musicians.[45]

Foreign bands were another source of competition the AFM worked to eliminate. Contracting with foreign orchestras, often for extended engagements, not only was less expensive than hiring union musicians, but the exotic names and music of such groups often had wide public appeal. Advertisements of the Royal Imperial Band of Wilna (Russia) as the "Special Favorite of the Czar," to cite an example, made the group sound far more interesting than familiar local bands. Local unions fought this practice by prohibiting members from performing for employers who imported foreign talent, but only the strongest locals challenged the practice successfully. The national union sought therefore to persuade Congress to extend the provisions of the Alien Contract Labor Law, passed in 1885, to musicians and musical groups. That law prohibited foreign workers with prearranged labor contracts from entering the country. However, it exempted "professional actors, artists, lecturers, or singers."[46] This exemption allowed contract musicians to enter the country and touched off a protest by American musicians that lasted fifty years.

Congressional debate over an 1899 tariff bill, which proposed a new tax on imported "implements of tradesmen," including musical instruments, provided the AFM an opportunity to renew its attacks on the policy of allowing foreign musicians to be contracted to work in the United States. The AFM demanded that Congress amend the contract labor law to cover foreign musicians. To fail to do so, union leaders said, would allow "men with instruments and gaudy uniforms on their back . . . to be classified as 'artists,' while simultaneously declaring that the materials these artists use are 'workingman's tools.'" Congress ignored AFM demands, and the inconsistency the union pointed to remained in the law until 1932. In belatedly making the change at that time, Congress finally accepted the union contention that musicians were workers first and artists second. After 1932 only "virtuosos of the first rank" were excluded from the restrictions of the contract labor law.[47]

The AFM also addressed the issue of peripatetic instrumentalists working outside the jurisdiction of their own union local. Unlike other craft guilds and protective societies, musicians' unions had always permitted their members to travel far and wide in pursuit of work. The 1872 constitution of the New York musicians' union, for example, authorized the issuance of "traveling cards" that extended the rights and protections of the union to members working in other jurisdictions, provided the traveling member paid dues to and abided by the rules of the local in whose jurisdiction he or she worked. It also extended the same privileges to members of other locals working in New York.[48] The AFM handled the issue of traveling musicians by instituting a "transfer law" that gave musicians the right to work in the jurisdiction of locals but allowed locals to prohibit the work of outsiders who came into their jurisdiction as a result of strikes, lockouts, or breaches of union contracts. In a nutshell, the union tried to keep musicians geographically dispersed without compromising their need or desire to travel.[49]

Racial matters were also a problem for the young AFM. Although some locals, such as Boston Local 9, accepted African Americans on an equal basis, many did not admit black musicians at all. The result was "colored" locals, the first of which seems to have appeared in Chicago in 1902, after a majority of whites in Local 10 voted for a segregated union. In response to the vote, black musicians quickly organized themselves and asked the AFM for a separate charter. According to William Everett Samuels, an early member of the black union, president Weber wanted to exclude black musicians from the federation. "[Weber] was so prejudiced," Samuels said, "that he didn't want [black musicians] either, but he couldn't keep them out, so he said all right, you [can] join the AFM but you'll be the colored local." The recognition of Local 208 set a precedent, and over the ensuing two decades separate black locals appeared in approximately fifty cities, including Cincinnati, Philadelphia, Pittsburgh, and Washington, as well as most southern cities. Some of these locals, however, were not autonomous. Until 1944 many operated under subsidiary charters from white locals.[50]

"King" Oliver's Jazz Band, Chicago, circa 1922. Oliver's was one of many professional groups that introduced and popularized jazz in the first two decades of the century. *(Bettmann Archive)*

Not all segregated locals resulted from the exclusion of black musicians from white locals. Many African Americans preferred their own organizations to those dominated by whites. In 1915, for example, several black musicians withdrew from Boston Local 9 on their own initiative and organized what became Local 535. Black musicians, explained one of the organizers, "wanted to have their own identity." No doubt they also wanted to ensure their own control over nightclubs and other workplaces catering to black customers. In any case, Local 535 was as aggressive as its white counterpart in asserting and defending the interests of its members. At a time when the demand for leisure activities was rising more rapidly than the supply of skilled instrumentalists, black as well as white musicians benefited. Despite a certain instability in its formative years, Local 535 protected wages and working conditions for black musicians in such popular Boston nightspots as the Royal Palms, Little Dixie, Louie's Lounge, the High Hat, Handy's Grille, the Old Savoy, and the Paradise Café. As one of its members, Ernie Trotman, said later, Local 535 "had certain kinds of work tied up."[51]

THE CHALLENGES MUSICIANS and their unions faced around the turn of the century contrasted sharply with those confronted by most other workers and unions. In that era of rapid industrialization and mechanization of manufacturing processes, skilled workers in many trades lost power and privileges in the workplace. New machinery divided previously skilled jobs into simple, repetitive tasks that semiskilled or even unskilled workers could perform, and in the process reduced not only skill levels and wages but workers' autonomy and bargaining power. Efforts to resist these developments and the new management techniques they spawned were more often than not futile, but the very futility of the efforts forced desperate workers across the nation, in eastern factories as well as western mines, into deadly acts of protest and resistance.

Musicians and their unions escaped these kinds of displacement and desperation in the late nineteenth and early twentieth centuries. They faced no innovative job-threatening machinery, no strong employer associations, and no efficiency experts speeding up the pace of work. Their union therefore was distinguished by its successes. It protected wages and income and facilitated the expansion of job opportunities. Although white males dominated the union, other groups carved out their own places in music and found ways to benefit from national organization. This was the Golden Age, the "good old days," which musicians and their union later looked back upon nostalgically.

Two

Boom and Bust in Early Movie Theaters

THE FIRST QUARTER of the twentieth century was the heyday of American musicians. Demand for musical workers was high and rising while the supply of skilled instrumentalists was relatively low. The public wanted and could afford entertainment with a large component of live music. Technological advances were also generally friendly to musicians. Phonographs, silent movies, and radio increased public appreciation of music and boosted employment opportunities. Electric streetcars, automobiles, and air conditioning brought more and more Americans to places featuring live music. The American Federation of Musicians (AFM) controlled the workplace and protected the interests of musicians there.

This state of affairs took a sudden, negative turn in the late 1920s, when the advent of sound movies helped transform the music sector of the economy into a more centralized, capital-intensive structure dominated by large business enterprises. Sound films "silenced" musicians as quickly as they ended the careers of silent-screen stars who spoke poorly. The "talkies" enabled theater owners to discharge pit musicians in wholesale fashion, a classic case of substituting capital for labor. By 1934 about twenty thousand theater musicians—perhaps a quarter of the nation's professional instrumentalists and half of those who were fully employed—had lost their jobs. The fact that this technological shake-up coincided with the onset of the Great Depression added to the woes of musicians, most of whom had few skills other than their musical abilities.

Instrumentalists did not stand passively by while capitalist development destroyed a major source of employment. Through their unions they waged a multipronged campaign to save theater jobs. To many Americans, especially theater owners, the struggle to save pit music was pure Luddism, blind opposition to technological progress. To professional instrumentalists, however, this was a fight not only to protect their livelihoods but to preserve their art and their dignity, and to maintain a measure of control over their own employment.

SILENT FILMS made their debut at a vaudeville theater in New York City in April 1896. At the end of a program of variety acts, flickering images projected onto a canvas screen amazed a theater audience, some of whom reportedly ducked when they saw waves rolling toward Manhattan Beach. Only a few years later several hundred vaudeville theaters advertised "moving pictures" along with comedy acts, dance shows, and other routines. The movies had quickly become one of the nation's most influential mediums of entertainment and culture.[1]

Invariably, theater musicians provided live music to enhance the effect of these early films. When vaudeville dominated the entertainment business, as it did at the turn of the century, most vaudeville theaters employed small in-house orchestras to enliven their stage shows. A typical house orchestra included five or six musicians. Quintets usually included a pianist, trap drummer, and violinist as well as a cornet and a trombone player. Larger orchestras might feature a clarinet, bass violin, flute, banjo, or organ as well. While silent films played, house orchestras tried to provide the appropriate music: dissonant chords and tremolos when villains plotted, soft violin music during romantic scenes. Drummers bumped the bass drum, crashed symbols, and played long rolls to add comic relief, tumult, and suspense. There were of course incongruities. One early film commentator accused local musicians of mangling movie scenes. "How often was the pleasure of seeing a stately military picture marred by the playing of a waltz or a ragtime selection," he asked, "or the picture of some pathetic scene, by the playing of 'Steamboat Bill'[?]"[2]

By 1905 growing numbers of entrepreneurs were converting pawnshops, cigar stores, and other such places into "nickel" theaters that showed movies from early morning until late at night. Some vaudeville, burlesque, and legitimate theaters turned to all-movie formats, at least a few days of the week. Film historians have estimated that in 1910 ten thousand theaters used movies as the core of their entertainment. Some of these houses hired ensembles to accompany films; others hired a pianist and/or drummer. Still others relied only on the music of player pianos. Pioneered by John McTammany of Massachusetts and perfected by William B. Tremaine of New York, player pianos were as new as movies themselves. Plugged into electric circuits, these novel instruments pumped air through strips of perforated paper (music rolls), which activated the keys of the piano. The music produced was no doubt better than that of some theater orchestras, for many prominent musicians recorded music for player pianos. Nonetheless, the low volume capacity of the instruments and the fact that they played only a few songs over and over, without regard for the character of films, made them undesirable to many theater owners.[3]

By 1910 piano manufacturers such as Rudolf Wurlitzer and J. P. Seeburg were producing more versatile automatic instruments especially for movie theaters. Models known as photoplayers proved the most popular. Priced from $1,000 to $5,000, photoplayers contained several music rolls and could play thirty or more songs without repetition. More important, photoplayer operators could switch from one music roll to another to create particular moods. "If the scene is a sad one and you want sob music," one manufacturer explained, "all you have to do is touch a button for the roll containing sob music." Side chests, meanwhile, contained an assortment of bells, horns, and percussion devices that could be activated by pushing pedals or by pulling straps located above and below the piano keys. The instruments thus produced the sounds of doorbells, fire trucks, auto horns, galloping horses, pistol shots, and various other noises. The more popular photoplayers had organ pipes in side chests that were activated by a second keyboard on the central unit. Organ pipes added volume as well as versatility to theater music.[4]

Some theaters used photoplayers as substitutes for live music. "We simply turn on the current in the morning and shut it off at night and the instrument does the rest," the manager of the Grand Theatre in Atlanta explained in 1915. Theaters that wanted to maximize the effectiveness of the photoplayers, however, hired skilled pianists to operate them. Most piano manufacturers apparently had the concerns of skilled pianists in mind when they designed the instruments. "Any good pianist with a little practice can play this instrument and produce all the various changes to suit the various shifting scenes of the pictures," one Wurlitzer advertisement explained. An advertisement for a popular Seeburg model promised to make "any pianist an organist" as well as "the master of every situation." The number of theaters that used photoplayers is unclear, but one study of

Player piano, 1909. These instruments could be either operated automatically (note the roll of music at the center) or played, and they became the chief means by which theater owners filled the void during the early silent-film era. *(Bettmann Archive)*

early movie theaters estimates that piano manufacturers sold six thousand to eight thousand of the instruments in the United States between 1912 and 1930. Photoplayers, however, had an average life span of only seven years and apparently broke down frequently. The fact that the instruments were expensive as well as unreliable no doubt encouraged many theaters to rely on live music for film accompaniment.[5]

In fact, more and more musicians were finding jobs in theaters. Changes in the structure of industry largely explain the trend. In the year preceding World War I, increasing concentration and centralization characterized the film industry. In the exhibition sector, entrepreneurs such as Marcus Loew were building networks of large theaters to capitalize on the public's interest in movies. Many of these theaters seated up to fifteen hundred people and featured vaudeville acts as well as silent films. In these houses live orchestras alone had the volume power and musical versatility to entertain audiences. Developments in the Midwest exemplified national patterns. In Milwaukee the nine-hundred-seat Princess Theater opened in 1909 with an eight-piece orchestra; the fifteen-hundred-seat Butterfly Theater opened in 1911 with a ten-piece band. In Chicago three new theater circuits emerged in the mid-1910s; collectively they owned or leased twenty theaters of eight hundred to fifteen hundred seats, and each of these theaters featured five- to eight-piece orchestras. Some of the houses were among the most lavishly adorned buildings in the city and, unlike many smaller theaters, appealed specifically to middle-class Americans.[6]

In the postwar years the drive for economies of scale encouraged the construction of still larger and more luxuriant theaters. By 1927, when Paramount, First National, Loew's, and Fox were fully integrated firms with heavy investments in film production, distribution, and exhibition, nearly one hundred theaters nationwide could seat more than twenty-eight hundred people. The rise of large movie palaces, to which some people reportedly paid the price of admission "just to use the restroom," increased competition for audiences, and the heightened competition meant bigger orchestras. In New York in 1927 the Capital Theater increased its orchestra to eighty pieces, and the Roxy advertised an orchestra of more than a hundred pieces.[7]

Other innovations similarly benefited musicians. In Chicago, for example, the Granada Theater put on a "Northwestern Night," for which it hired the house orchestra to play for two hours in the lobby after the final show while students from Northwestern University danced. Theater owners also tried to capitalize on the jazz craze then sweeping the nation, and

Piccadilly Theater, Chicago, 1927. Orchestra pits in period theaters like this one could be raised or lowered as the occasion demanded. The Piccadilly's pit organist used remote-control mechanisms to play the piano and harp adorning the two balconies nearest the stage. *(Theatre Historical Society)*

some of those who had eliminated vaudeville acts because of the popularity of movies rehired them, and with them more musicians. Equally important was the opening of the world's first mechanically cooled theater, Balaban & Katz's Central Park Theater in Chicago, in 1917. The advent of air conditioning allowed exhibitors to keep their theaters open all year, and thus served to increase musicians' job opportunities.[8]

These were indeed opportune times for musicians. The expansion of theaters meant steady, well-paid work, and in some locales the demand for house musicians soon exceeded the supply. By 1928, when approximately twenty-eight thousand theaters blanketed the nation, upwards of twenty-five thousand musicians worked in front of silent screens. Theaters in New York City alone supported thirty-two hundred musicians.[9] AFM president Joseph N. Weber estimated that theaters offered more full-time job op-

Stage to Studio

Table 2 AFM Membership in Selected Locals, 1918–1928

City and Local	1918	1920	1922	1924	1926	1928	Growth during Period (%)
Cleveland (4)	868	1,014	1,639	1,311	1,412	1,458	68
San Francisco (6)	1,250	1,600	1,800	2,350	2,425	2,700	116
Boston (9)	1,716	1,835	1,901	2,143	2,250	2,459	43
Chicago (10)	2,850	3,166	3,943	4,256	5,728	7,146	151
Newark (16)	600	724	860	1,058	1,293	1,442	140
Kansas City (34)	525	704	706	765	845	952	81
Baltimore (40)	668	806	1,110	1,152	1,244	1,239	85
Omaha (70)	271	421	456	514	529	581	114
Memphis (71)	153	149	154	186	255	250	63
Minneapolis (73)	833	1,010	1,264	1,102	1,049	1,148	38
Seattle (76)	564	867	924	1,085	1,222	1,388	146
Atlanta (148)	170	232	273	314	346	333	96

Source: Official Proceedings, 1918–28.

Note: This table sheds light on the steady growth of AFM membership between 1918 and 1928. The growing size of AFM locals around the country was largely a consequence of expanding job opportunities.

portunities than all other sources of musical employment combined: dance halls, hotels, symphony orchestras, restaurants, and cafés. Weber noted too that wages of musicians had doubled in the past twenty years, a fact that more than compensated for postwar inflationary trends.[10]

Such patterns reflected the growing power of the AFM. Membership in the union had nearly doubled since 1918, from about 80,000 to more than 150,000 (perhaps half of whom were "amateurs"), divided into 780 locals representing every city of any size in the country. Using practical bread-and-butter tactics, many locals achieved what amounted to monopolies on musical services. They forced theater owners, who had no parallel organization of their own, to hire union members only. According to union sources, 98 percent of American theaters had closed-shop contracts with AFM locals. Because theater owners suffered irretrievable losses when musicians went on strike, the mere threat of a walkout could typically force them to agree to union demands.[11] The AFM, then, had taken advantage of a favorable setting to assume price- and market-regulating functions.

THE WORK SCHEDULE of theater musicians varied according to several factors. In the early 1920s most theater musicians performed seven days a week during seasons that ranged from thirty to fifty-two weeks, depending on location. In balmy Southern California, musicians generally worked year-round, but in Columbus, where some theaters apparently were not air-conditioned, they worked thirty weeks and then negotiated extra engagements during the summer months.[12] Musicians performed between four and seven hours a day, with the time typically being divided between one or two evening shows and perhaps an afternoon matinée (many theaters hired only a single pianist or organist for matinées and late-evening shows). The time instrumentalists actually worked, however, was longer than these numbers indicate. Theater musicians usually rehearsed for each new film or vaudeville show, without compensation. But beyond that, rehearsal time meant extra wages. Musicians in Boston, to illustrate the pattern, earned overtime wages when daily performances exceeded five and a half hours, and when they had to rehearse on Sundays.[13]

Long workweeks sometimes sparked protests. Complaining of having their "nose[s] to the grindstone" 365 days a year, San Francisco musicians in 1926 demanded that theater owners hire "capable" substitutes one day a week. The demand spread to Los Angeles, where theater musicians asked to be able "to live like other human beings." "Six days shalt thou labor," one of them exclaimed. Drawing on this militancy, musicians in these two cities withdrew their services from the Orpheum and Pantages theater chains in September. The owners immediately agreed to six-day workweeks and meaningful salary increases. The swift capitulation reflected the importance as well as the limited supply of qualified instrumentalists.[14]

A recent interview with Gaylord Carter, an organist who in 1922 moved to Los Angeles from Wichita, Kansas, reveals the nature of theater employment patterns at this time. Carter found his first job in Los Angeles at age seventeen when the owner of a local theater asked Carter's father if he knew anyone who could play the organ for silent movies. Carter said he took the job primarily to see the movies. "I didn't have a dime to get in to see the shows," he recalled, "so I got a job playing [organ] in the theater." In 1926 Carter moved to the much larger Million Dollar Theater, which featured a thirty-six-piece orchestra. There he worked six days a week because, he said, "the union required that you have one day off." As Carter explained things, orchestra members staggered their days off so that there would be "a few guys off each day." Other Los Angeles musicians followed this pattern of employment.[15]

Seated in the orchestra pit in front of the stage, theater musicians worked the entire period of each show. They performed diverse repertoires,

Table 3 Wages and Rules of Theater Musicians in Columbus, Ohio, 1921

WAGES FOR THEATER WORK:

First Class Theaters: (Admissions exceed $1.00)
Per Man, per week of six days . $39.00
Leader, per week of six days . . $58.50
Sunday shows, per man $7.50
Sunday shows, leader $11.25

Second Class Theaters: (Admissions do not exceed $1.00)
Per Man, per week of six days . $39.00
Leader, per week of six days . . $58.50
Per Man, per week of seven days . . $45.50
Leader, per week of seven days . . $68.25

SELECT RULES GOVERNING THEATER WORK:

1. The house Leader shall have full charge of the men, to engage or discharge them and shall receive the entire Leader's salary.

2. One two-hour rehearsal gratis for each engagement, extra rehearsal, morning or afternoon, 2 and 1/2 hours or less per man, $4.00; Leader, $6.00. Night rehearsals, 3 hours or less, per man $6.00; Leader, $9.00.

3. Contracts for the Summer season must be for a period of ten consecutive weeks or more.

4. The minimum number of men law applies to all Theaters and Picture Houses and Halls in the City District, according to the highest price of general admission as follows:

$1.00 and higher . . . 8 men
50 to 99 cents . . . 7 men
40 to 45 cents . . . 6 men
30 to 35 cents . . . 5 men
20 to 25 cents . . . 4 men
10 to 15 cents . . . 3 men

5. Substitutes in all Theaters shall receive 50 cents extra per show more than regular men up to a full week's salary. Extra men shall receive $7.50 per day, one or two shows.

Source: Price List of the American Federation of Musicians, Local 103, MIC 155, vol. 11 (1921), 1–5, Microfilm Department, Archives–Library Division, Ohio Historical Society, Columbus.

Note: These rules, adopted by the Columbus musicians' union in 1921, illustrate the nature of the protection that unions provided theater musicians. Musicians in Columbus divided theaters into three categories according to the price of admission. The wage rates presented here applied to first- and second-class theaters for a season of thirty weeks. The rates applied to men and women equally.

Aldine Theater Orchestra, Pittsburgh, 1928. Theater orchestras accompanied silent films and variety shows and often performed on stage. Here comedian Bennie Rueben, a vaudeville favorite, stands to the left of the musicians. *(Bennie Rueben Collection, Cinema-Television Library, University of Southern California)*

generally opening with classics like Chopin's "Nocturne Number 5" or Schumann's "Sunday on the Rhine." When variety acts took the stage, the orchestras supported them. For singing comics they played novelty songs, for dancers perhaps a ragtime tune. After five or six vaudeville acts the orchestra again played classics, or perhaps a set of popular tunes like "A Trip to Coney Island," "Thanks for the Buggy Ride," or "Rhapsody in Blue." To create proper moods, orchestra leaders drew variously on their assortment of clarinets, flutes, saxophones, trumpets, trombones, tubas, violins, pianos, and drums.[16]

By 1920 motion-picture studios were providing musical scores, called "cue sheets," for their films. Producers hoped thereby to determine, or at least influence, the music that accompanied the screening of their films. A typical cue might call for a specific minuet by Haydn "for ninety seconds until title on screen," or for a piece by Tchaikovsky "for two minutes and ten seconds . . . until scene of hero leaving room." To musicians such cue sheets were "mutilated masterpieces." One music publisher who provided cue sheets to film companies admitted that his employees, usually men with limited composition skills, simply cut up classics to fit film scenes. "We murdered everything that wasn't protected by copyright," he said. For this reason many exhibitors refused to buy or rent cue sheets, and even when they did, orchestra leaders often ignored them. One reason for this was that "local ego" sometimes clashed with the proffered musical scores. According to one theater musician, conductors sometimes told theater managers of the scores, "Anything we could do would be better."[17]

In all of the music accompanying silent films, the sounds of the house organ were most recognizable. The introduction of Wurlitzer's large Hope-Jones Unit Orchestra in the mid-1910s had significantly increased the value of organs to theaters. One difference between the "Mighty Wurlitzer" and earlier organs was the Wurlitzer's better utilization of air pressure, which made for much more brilliant musical tones as well as greater volume capacity. A system of "pipe unification" also allowed organists to trigger many pipes at once with the touch of a finger; in contrast, the old system of rope and knob pulling had been able to activate only a few pipes at a time. More important, because Wurlitzer shaped organ pipes to reproduce the tones of particular instruments, organists had at their command the sounds of a full orchestra. The dozens of colored stop tabs arranged in horseshoe fashion across the instrument's console were labeled violin, cello, flute, tuba, oboe, piano, and the like. Compared with the Wurlitzer, then, other organs sounded "sacred" at best and dull at worst.[18]

There were still other reasons for the name Mighty Wurlitzer. The versatility of this organ revolutionized sound effects for silent films. The instrument could create not only old sounds like steamboat whistles, quacking ducks, and gunshots but much more nuanced moods as well. Moreover, an advanced electro-pneumatic relay system liberated Wurlitzer consoles from direct physical connection with organ pipes. As a result, consoles could be raised and lowered from orchestra pits, much to the delight of audiences. The ascent of spotlighted console and organist became a celebrated part of the show in theaters that installed the necessary lift. The willingness of the audience to suspend disbelief, so essential to the success of the film, typically rose along with the Wurlitzer. For all of these reasons, growing numbers of theaters purchased Wurlitzers, and other organ manufacturers began producing their own versions of the instrument, including the Robert-Morgan, Kimball, Kilgen, Moller, and Marr and Colton companies.[19]

The impact of the new organs on theater employment patterns is unclear, but several factors prevented the instruments from displacing musi-

A Kimball organ, circa 1927. Such versatile theater organs brought a new sense of excitement to movie houses during the 1920s. *(Museum of Modern Art)*

cians on a grand scale. Many theaters could not afford the new organs, which cost from $20,000 to $40,000. Those that could afford the organs sometimes had difficulty finding organists who could play them, especially in the 1910s. The complexity of the instruments required organists to demonstrate a distinctive combination of physical dexterity, musical skill, and even mechanical ability. The expanding number of theaters in the postwar years only added to this problem of labor scarcity. Then, too, the best organists in the country were typically members of the AFM, and as such they complied with union rules governing the size of house orchestras as well as the wages and working conditions of orchestra members.[20] Not surprisingly, the men and women who played theater organs well enjoyed full employment, high wages, and job security.

The recollections of Helen Lee, who played at the Milford Theater in Chicago in 1925, ellucidate the work routine and status of theater organists. Twenty-one at the time, Lee worked for seven nights a week in the pit alongside, but independent of, a seven-piece orchestra. From one to five P.M. she taught music at the Chicago Musical College, after which she ate dinner, took a taxi to the theater, and prepared for a seven o'clock per-

Stage to Studio

formance. "I got a bite to eat," she recalled, "and went straight to the theater." There she dressed and waited for the show in her own dressing room, a perquisite none of the orchestra players enjoyed. "There were several rooms for actors and vaudeville people," she said. "My room had shelves, a washroom, and my music library." From the library Lee selected the music she would play for the evening's silent movie. "The picture usually came with a cue sheet," she said, "but we changed the music." The theater manager, for whom Lee worked, did not always approve of her changes. One night she substituted Ravel's "Pavane on the Death of a Royal Infant" during the child murder scene in *King of Kings: The Life Story of Jesus*. "The audience doesn't understand that kind of music," the manager told her after the movie; "you'll have to lower your standards." Her reply reflected the independence as well as the pride theater organists had in their work: "I'm sorry, but I can't lower my standards. The audience will have to raise theirs."[21]

Lee's instrument was an elegant $35,000 Kilgen organ, which she preferred to the Wurlitzer. "I loved the Wurlitzer," she recalled, "but the Kilgen was more mellow." Playing the Kilgen, Lee enlivened variety acts as well as silent films. She performed both solo and with the house orchestra. During the course of an evening, she said later, "I played more than they did." In fact, she and they spelled each other during the show. A union member, Lee earned union wages, $100 a week, which was more than the orchestra musicians earned, and far more than the average skilled worker. "Men with families were earning twenty dollars a week," she said, "and I made a hundred."[22]

Although many female musicians worked in theaters, especially as pianists and organists, theater work posed special challenges to women. "It was dangerous," Lee said, "to come home late at night." Because of the danger, friends and family often discouraged women from working in theaters. "My parents were always hoping that something would happen to make me lose my job, they always worried," Lee explained. Like many other women, however, Lee accepted the risks and took her work seriously. On occasion, for example, she remained at the theater until four A.M. overseeing the work of organ tuners "to be sure they did it right."[23]

Excluding the stars on the screen, only orchestra leaders rivaled house organists in popularity. This was partly due to the fact that orchestra leaders did much more than simply cue the musicians they directed. Many of them served as masters of ceremonies, announcing vaudeville acts, introducing movies, and hobnobbing with performers on stage. At the Orpheum in Los Angeles, one especially hardworking leader had "to kiss every girl in the act" while the band played the popular song "Gilded Kisses."[24] The best-known leaders were themselves main attractions.

Theater managers contracted with the leaders for the services of the band, and the leaders paid themselves at least 50 percent more than they paid the sidemen, whose services they had subcontracted. Union rules not only required higher wages for leaders but allowed them considerable authority in the pit. Leaders could fine or dismiss sidemen, order rehearsals, and prescribe dress codes. The union, however, offered sidemen protection against overly demanding leaders. It required additional pay for long rehearsals and special costumes, for example, and insisted that leaders make sure orchestra members had refreshments. In regard to dress in the pits, no general rule prevailed. Some leaders required tuxedos, while others permitted more casual wear. One reviewer of film and vaudeville in Los Angeles noted that in some theaters musicians "dressed up like a movie star's poodle" while in others they wore "everyday pants" and "no collars."[25]

As sidemen looked to orchestra leaders for cues as well as paychecks, leaders received their direction from theater managers. Relations were usually cordial, but when managers meddled or tried to meddle in matters leaders thought were their own responsibility or prerogative, tensions might surface. Leaders and sidemen sometimes viewed managers as audiences viewed villains on the screen. *Theater Magazine* expressed the feelings of those who did when it described one theater manager as "a man who conducts his business in bursts of emotion and rarely with any other ambition than to earn a fortune quickly." Managers generally, according to the magazine, were "sub-average or ab-average," seldom having "any foresight." The trade paper of Los Angeles musicians described managers as both "aggressive" and "stupid," men "plucked from the ranks of salesmen and ushers." One manager, the paper noted, "has been known to seat himself in the last row of the balcony and scrutinize the orchestra through an opera glass, trying no doubt, to see if the tuning pegs on the violin were all set in the same angle."[26]

Whatever tensions existed between musicians and managers, the popularity of movies and the competition among theaters boosted musicians' wages and employment opportunities. By 1928, when approximately a quarter of all professional instrumentalists worked in theaters, monthly wages of sidemen in the most competitive places reached $300, while organists in such locales might make $400. By comparison, skilled workers in building trades generally earned less than $150 a month throughout the 1920s. The status of theater musicians never rivaled that of touring concert artists or musicians in symphony orchestras, but pit musicians, especially organists, enjoyed immense popularity.[27]

THIS SITUATION began to change as soon as entrepreneurs brought "canned music" to the movies. Al Jolson's warning, "You ain't heard nothing yet" (*The Jazz Singer*, 1927), is often called the beginning of the sound era in movies, but this is not strictly correct. The 107-piece New York Philharmonic Orchestra had earlier recorded an accompanying score for *Don Juan*, a 1926 Warners film starring John Barrymore and Myrna Loy. Using new Vitaphone sound equipment, exhibitors now matched discs of recorded sounds—voices, special effects, and music—to scenes on the screen. Although the sounds were scratchy, critics sensed at once that *Don Juan* and *The Jazz Singer* opened "a new era in motion pictures." If the full implications of these movies were unclear at first, entrepreneurs soon realized that the talkies would rid them of the high cost—and trouble—of maintaining theater orchestras and employing vaudeville acts.[28]

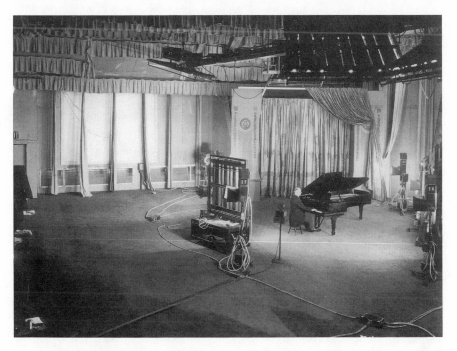

Fox Chase Corporation sound studios, New York, circa 1927. These studios produced some of the nation's first sound films. *(Case Research Lab Museum)*

But sound movies did not spread instantly or automatically. On the contrary, investment in traditional production methods discouraged use of the new technology. Film moguls saw at once that if the public demanded sound movies, they would have to allocate huge sums of capital to remodel production studios as well as movie theaters. Talking pictures also threatened the careers of silent-film stars, who were already under contract and whose popularity studio publicity departments had carefully crafted. The resulting resistance to sound demonstrates that management as well as labor could oppose technological change. Just as workers feared that innovation would mean loss of jobs, wages, and status, so management worried that new processes would render profitable investments obsolete.

The fact that Warner Bros. pioneered sound movies is a datum that speaks to this pattern of reaction to technological innovation. When the first talkies appeared, Warners was a small company in a weak competitive position, but it had sufficient resources and marketing power to pursue bold, expansionary strategies. The company paled in comparison with industry leaders like Loew's and First National, but with the backing of a Wall Street investment house that specialized in turning promising regional businesses into national enterprises, Warners built a vertically integrated firm that not only produced and distributed films but screened them in its own nationwide chain of theaters. As it expanded aggressively, Warners gambled on Western Electric's new methods of reproducing sound, and like many other firms in a similar position, it supported technological innovation for the specific purpose of minimizing labor costs. Simply put, the firm decided to substitute cheap recorded sound for the live sound that pit musicians had provided for silent films. "I saw the salvation of the cinemas," Harry Warner reportedly remarked, recalling his motives for making the investment, "[in] the defeat of the vaudeville invasion that was seeking to dominate the cinema theaters."[29]

The savings the new sound technology promised theater owners were considerable. The annual cost of a sixteen-piece orchestra was perhaps $50,000, while one of the new sound systems cost from $7,000 to $25,000, depending on the type of system and the theater's seating capacity. Owners of small theaters often had difficulty raising the money necessary to make the change, but the potential savings in labor costs were substantial. One historian estimates that exhibitors saved as much as $3,000 a week by displacing musicians and vaudeville actors. Another calculates that when exhibitors installed sound systems, their net profits rose as much as 25 percent.[30]

But substitution of capital for labor is seldom just a matter of money. Despite occasional breakdowns and problems of synchronization, sound technology was far more reliable than actors and musicians. Talking movies did not demand higher wages, go on strike, or fail to show up for work. Nor did they argue over song selection. Sound movies thus brought rationality to theater operations and made the task of management easier. Such advantages of machinery over human labor always encourage technological innovation, especially when it increases profits, as it did in this case.

The public's response to the new technology removed all doubts about the practicality of conversion. Audiences stood in long lines to see talking pictures; box office receipts skyrocketed. Warners' investment in Vitaphone soon lifted the company to the top of what had become an extraordinarily competitive industry. Assets of the firm rose from $5 million to $160 million during the last two years of the 1920s. The Fox Film Corporation and Radio-Keith-Orpheum (RKO) also benefited enormously from the conversion to sound. Movietone, the sound-on-film method of registering and reproducing sound waves developed by the two latter companies, proved to be far less cumbersome than Warners' sound-on-disc system. In June 1929, technicians were installing Movietone systems in theaters at a rate of fifty a day. By the end of the year four thousand theaters had sound reproduction systems or were in the process of installing them. Movie attendance climbed from fifty million a week in 1926 to ninety million in 1930, or an average of almost one visit a week for every American.[31]

While some moviegoers may have missed the sounds of live orchestras, the vast majority, as the attendance figures suggest, preferred the new system. The quality of music in sound movies often surpassed that of local musicians, especially in small towns. The quality of film music, however, did not by itself explain the burgeoning size of theater audiences. Sound technology brought new excitement to the screen, invigorating individual film genres. The sounds of real guns firing, glass breaking, and tires screeching, for example, brought new life to gangster films. Musicals also became highly popular once producers learned now to synchronize recorded music with dancing and singing on the screen. In short, sound movies were novel, and novelty has always had its own value in the marketplace.

Advertising also spurred the popularity of sound movies. Producers and exhibitors who invested heavily in the new technology spent lavishly to advertise its merits. Studio publicity firms had little difficulty convincing the public that talking movies represented a new level of entertainment as well as applied science and were thus the only acceptable form of motion-picture art. As Fox studios explained, sound movies were "the result of half a century's experience in telephone making." Advertisements portrayed silent films, at least indirectly, as obsolete and those who preferred them as eccentric. Moviegoers, however, needed little convincing; they were as eager to hear as they were to see their favorite movie stars.[32]

FILM HISTORIANS variously describe the transition to sound as a "coup" and as a "revolution." Both terms reflect the extent to which technological change usurped the power of labor and enhanced the prerogatives of management. The advent of sound movies meant crisis for vaudevillians and theater musicians, whose source of employment collapsed just as the Great Depression began. On October 28, 1929, the day before the stock market crash, *Film Daily* reported that nearly a third of the nation's theater musicians were already jobless.[33] In the aftermath of the crash, the larger depression reinforced the effects of technological displacement, and both vaudevillians and theater musicians joined the ranks of dinosaurs, dodo birds, and other extinct species.

The situation in Chicago illustrates the pace and impact of these changes. In 1926, musicians there had a secure position in movie and vaudeville theaters. In September of that year they won meaningful wage increases in negotiations with nine Shubert theaters employing nearly four hundred musicians. J. J. Shubert himself flew in from New York to sign the new contract, which raised weekly wages almost 10 percent for regular orchestra members (from $72.50 to $79) and doubled rates for rehearsals (from $2 to $4 a session). Encouraged by this success, the union struck smaller theaters, confident that management realized that "movies without music are not popular." Within four days the intimidated exhibitors' association had come to terms, accepting a three-year contract that boosted weekly wages for musicians from $82.50 to $87.50, doubled the pay for rehearsals, and obligated the city's smallest ("Class No. 6") theaters to hire four-piece orchestras. In the wake of this victory more than twenty-five hundred musicians worked in theaters in Chicago, and the president of the union local there, James C. Petrillo, stated, "The musicians' union is stronger than ever. We are certain that the theater managers will never again attempt to put anything over on us."[34]

Over the next two years Chicago musicians lost ground steadily, and Petrillo's tone moderated. In July 1928, after thirty theaters had installed sound systems, Petrillo voiced the "alarm" and "fear" that now gripped musicians. Publicly he insisted that "mechanized" theater music was only a fad, and that audiences would always demand to see as well as hear musicians. Privately he was less confident. In September 1928 he warned that if theater owners ignored the interests of musicians, "the question of declaring open war against mechanical devices in general can then be considered and, if need be, put into execution."[35]

The sudden threat to theater jobs stunned AFM leaders and musicians alike. Ever since the advent of recorded music, the leaders had watched the progress of sound technology and denied, publicly at least, that machines would ever replace artists. Even after sound movies appeared, they insisted that interest in "mechanical music" would prove temporary. Theater musicians reacted similarly. "We all thought it was just a fad," organist Gaylord Carter said later; "we thought it would pass." A prominent member of the Los Angeles Theater Organists' Club predicted of talkies that exhibitors would "lose their shirts in this latest folly" and see their theaters "turned into parking lots."[36] An editorial in the trade journal of the Los Angeles local predicted that talkies would fade because they forced ocular and auditory nerves to "pull in double harness." "Relaxation," the journal explained, "is thereby decreased by 50 percent." Other musicians agreed; machines, they insisted, would never produce "that illusive something" that live orchestras alone provided "unless the secret of life be discovered and its functioning controlled."[37]

The proliferation of theater sound systems in combination with new, higher-fidelity recording methods promised hard bargaining sessions in the 1928–29 season. Fearing for the livelihood of over twenty thousand of their members, delegates to the June 1928 AFM convention debated their options. Some suggested that theater owners be urged to increase admission fees, in order to continue to employ musicians (though the increase would also enable them to cover the cost of new sound equipment). Others proposed that musicians accept lower wages. Still others proposed that union members be banned from recording music, a move that would have eliminated hundreds of new jobs in the media centers of New York and Los Angeles and with them an important new source of employment. The proposed ban on recording had the support of many locals, including the one in Chicago. The national president, Joseph N. Weber, insisted, however, that the ban would not only fail but also "hold us up to . . . ridicule." Musicians, like other workers, would be unable to resist the advance of technology. "The development of machinery cannot be hindered," Weber declared; "there is no force on earth—or ever will be—able to do this."[38]

Instead of blindly opposing technological innovation, Weber argued, musicians should rally public support for live music in theaters. At his urging, the convention organized a "Theater Defense Fund" to finance a public "educational campaign." Independent of the strike fund, the new fund came from a 2 percent tax on the wages of theater musicians and would be used to tell the public that the uncontrolled use of recorded music was not "progress" but "debasement of music." In a complementary move to increase union activism on this and related issues, the convention also raised union dues to bring an additional $1.5 million a year into the national treasury.[39]

These actions came at the proverbial eleventh hour. Less than two months after the 1928 convention adjourned, labor contracts across the country expired, and because of the rapidly changing situation, exhibitors refused to meet union demands. In Milwaukee, St. Louis, San Francisco, and elsewhere, musicians were forced to accept wage cuts, reductions in the size of orchestras, and abbreviated employment seasons. In other places they struck, only to be completely defeated; in New Orleans, for example, two theaters replaced striking musicians with Vitaphone and Movietone sound systems, while in Michigan City three theaters responded to walkouts by discontinuing live music because of "an inability to come to terms . . . with the musician's union."[40]

Events in Chicago demonstrated the escalating problems. There, contracts with fifty theaters expired on September 3, 1928, and anticipating a musicians' strike, exhibitors had already hired "cue boys" to operate sound effects devices. In negotiations with the union, management insisted on reducing wages as well as orchestra sizes. When Petrillo threatened to strike, management obtained an injunction from Judge James H. Wilkerson prohibiting a strike against the city's largest theater chain, Balaban & Katz, and enjoining the union from "intimidating" theater owners and expelling members who continued to work without contracts. The labor press described the case as one in which exhibitors "went out the back door" and bribed a judge with an "unsavory reputation." In the aftermath, however, Clarence Darrow, counsel for the union, could only advise Local 10 not to renew theater contracts. No law, he said, could compel a person to work without a contract.[41]

The Chicago injunction was another indication that musicians faced many of the same problems as other skilled laborers in periods of industrial and technological transition. Court orders favoring employers and limiting trade unionists had been facts of life for American workers since 1880, but the practice became more widespread in the 1920s, when courts issued more than nine hundred injunctions against unionists. Backed by the courts, employers now challenged musicians and their unions as they had earlier challenged—and overwhelmed—railroad workers, miners, and other craftsmen and unions. Whether aimed at musicians or other workers, the injunctions showed that judges did not see management and labor as equal partners in a free-enterprise system. But concerning theater musicians specifically, judges no less than other Americans seemed to see them as obsolete and their efforts to hold on to jobs as Luddism.

When the Chicago labor contracts expired, more than seven hundred instrumentalists refused to work. Local 10 then asked other amusement trade workers for help, and stagehands and projectionists agreed to support the musicians. Exhibitors, many of whom were not yet prepared to declare their independence from live talent, settled with the musicians. This ability to muster support from other unions was a sign of Local 10's vitality, but it also indicated that the union could no longer fight its battles alone. The declining leverage of Local 10 was painfully apparent in the new contracts, in which the union accepted wage cuts, reductions in orchestra size, and in many theaters a shortened season as well. Moreover, the new contracts were for one year as opposed to the usual three.[42]

The next year, 1929, theater musicians from Charleston to Seattle suffered similar setbacks. Thousands lost their jobs; but despite the spectacle of mass displacement, the AFM continued the strategy of adapting to changing conditions. Opposing confrontational strategies, Weber continued to insist that the fate of theater musicians lay with consumers; the musicians could survive, he believed, only if consumers demanded live music. In early 1929 he therefore besought local officials to make no-strike pledges and undertake a nationwide campaign to sell the case for live music.[43]

Following Weber's lead, the national union adopted a "Declaration of Principles" and agreed to "spare no expense" in promoting the cause of theater musicians. The substitution of mechanical for live music, the union insisted, was "a perversion which constituted a fatal blow to musical culture," a step backward that would have a long-lasting detrimental effect on American culture. Unrestricted use of recorded music would not only deprive deserving musicians of jobs; it would also deaden public appreciation of music. More important, by eliminating the largest source of employment for musicians, "canned" music would discourage talented youths from considering musical careers. For its own sake the public should therefore demand "that the field for the creation of professional musicianship be not destroyed."[44]

This campaign reverberated through the labor press. In August 1928 the *Los Angeles Citizen* published an appeal from H. P. Moore, a local union official: "Substitution of mechanical music inevitably means a debasement of the art of music. Our national music will be seriously affected if 'canned music' . . . reduces the musicians' opportunities of employment. Where will the young musician of the future gain the incentive to perfect his art if a mere handful of recording artists are supplying all the music needed?"[45] A month earlier the Chicago local's *Federation News* had noted that more than two hundred theaters already had new sound systems, and it warned that the nation faced "a deplorable alteration of its musical entertainment."[46]

Over the next two years the AFM spent $1.2 million trying unsuccessfully to rally public support for theater musicians. Because of its simultaneous fight with broadcasters over the use of recorded music on radio, the union shortsightedly rejected the use of radio advertisements in this campaign. Instead it placed its ads in nearly eight hundred newspapers and twenty-five magazines. In cities where newspapers were especially anti-labor, the union rented billboard space. While some of the advertisements warned of the "anti-cultural activities" of theater owners, others sought to exploit antimodernist fears of mechanization. "The Robot as an Entertainer," read one advertisement: "Is His Substitution for Real Music a Success?" Underneath the question was a cartoon in which an iron man ripped out the strings of a harp while a dog howled and an angel wept. The advertisements asked the public to show support for live music by joining the "Music Defense League," which required only that readers sign and mail a printed coupon to the AFM. Union officials planned to use the signatures to create pressure to "keep music alive."[47]

Although an estimated three million people signed Music Defense League coupons, in rapidly increasing numbers the public lined up for sound movies. This escalating support for sound films reinforced the exhibitors' belief that their own survival in the marketplace depended on adopting the new technology. Even small theaters with small labor costs and few resources began to convert to sound. In 1930 hundreds of theaters released their orchestras and replaced them with new sound systems. In accomplishing this change, management typically refused to sign contracts requiring minimum-size orchestras, and when their new sound systems were in place they simply fired musicians. A few theaters continued to use organists as soloists between movies, but only until it became obvious that live music was unnecessary. In 1929 the president of the Exhibitor's Association of Chicago, the trade association of theater owners, summed up the situation: the new contracts, he said, "will permit us to hire and fire the men as they are needed."[48]

By 1930 the balance of power between exhibitors and musicians had shifted decisively to exhibitors. Management freed itself of musicians, answering strikes across the nation with "all-sound" presentations and lower admission fees. In Philadelphia, where Warner and Fox theaters demanded a 65 percent cut in orchestra size, the union withdrew the services of all of its musicians, only to find theaters continuing to operate with "no complaints from the public." When Publix theaters in Minneapolis and St. Paul introduced sound movies, management reported that public complaints were "practically nil."[49]

Musicians' problems multiplied after 1930, when the displacement of theater orchestras intersected with the much larger problem of the Great Depression. The film industry began to feel the depression just as the novelty of sound movies was fading. By 1932 weekly attendance had fallen from eighty million to sixty million, and it remained at the lower figure during 1933. At the same time, annual box office receipts dropped from $730 million to $500 million. The resulting decline in profits jeopardized the film industry itself. The recent expansion in the number of movie theaters combined with the continuing cost of sound conversion to leave many companies overextended and unable to meet their financial obligations. Warner Bros. lost $8 million in 1931 and another $14 million the following year, and Fox, RKO, and other major studios suffered similar losses. Paramount's financial troubles affected all of its employees, including musicians. Between 1931 and 1934 the company laid off five thousand workers, most of whom earned $35 to $50 a week.[50]

The drop in movie attendance also put enormous pressure on small independent exhibitors, who reacted in various ways. Some independents in the Midwest began offering "two seats for one," while others in New England introduced double features, a practice that soon swept the nation. Some exhibitors introduced games like bingo at intermission, with cash prizes for winners. Most theaters, however, simply lowered ticket prices. The drive to increase patronage went hand in hand with the determination to lower overhead costs, which included getting rid of musicians. Small theaters that had resisted conversion to sound now found that the alternative to sound was bankruptcy.[51]

The combined pressures of technological change and economic depression did not affect all business equally. Large national theater chains like

Balaban & Katz, despite their financial problems, had distinct competitive advantages over independent exhibitors. Small, usually family-run theaters generally had poor credit ratings as well as inferior managerial skills and were completely dependent on local market conditions. Confronted with local bank closures, reduced public spending, and the costs of conversion to sound, many of them closed their doors. A recent study sheds light on the trend. Between 1926 and 1937 the number of theaters in Chicago with fewer than 350 seats declined from 116 to 49, and the number with 350 to 1,000 seats declined from 173 to 160, but the number with more than 1,000 seats increased from 99 to 108. Nationwide the total number of theaters in operation in the 1930s declined by a third, from approximately twenty-three thousand to fifteen thousand, though as the trend in Chicago indicated, the ratio of people to seats remained fairly stable.[52]

Musicians and their union made desperate efforts to save theater jobs. They accepted lower wages, dropped demands for minimum-size orchestras, and agreed to restrictions on working conditions, but all to no avail. By the summer of 1931 approximately half of the nation's theater musicians had lost their jobs.[53] In New York City seventeen hundred musicians' jobs, 53 percent of all theater employment there, disappeared between 1929 and 1931. In Chicago the employment of theater musicians dwindled from nearly 2,000 in the late 1920s to only 125 in the mid-1930s. By 1934 only forty-one hundred theater musicians were still employed nationwide, and many of these lost their jobs during the next few years.[54]

Conditions in Washington, D.C., reflected these patterns. There, in August 1930 the Motion Picture Theater Owners' Association informed AFM Local 161 that orchestras in downtown movie houses would be dismissed in a month, when their contracts expired. "Two hundred men who have spent their lives in perfecting their art," one labor paper said of the dismissal, "will be scrapped"; and A. C. Hayden, president of the local union, found that theater owners "would not discuss the making of a contract to employ even one man." Their plan, Hayden said, "is to get rid of the musicians," who as a result will "be forced to compete for jobs in other industries for which they are not trained."[55]

In 1930 forty-eight theaters in the nation's capital employed 193 musicians; after the signing of new labor contracts, the number dropped to 72. Only three downtown deluxe houses still had orchestras, and even they had downsized their orchestras from a total of seventy-eight to sixty-one musicians. The remaining forty-five theaters had laid off 104 of 115 musicians. Over 60 percent of the city's theater musicians lost their jobs in this brief period. Of 101 "technological casualties," a survey found that 20 had left the city; 11 had full-time and 19 had part-time musical jobs; 21 had jobs in nonmusical fields; and 22 were unemployed. The fate of eight others was unknown.[56] In making this survey a representative of the Labor Bureau visited fifty displaced musicians in their homes and found them in generally low spirits. Many reported that the experience of displacement itself was a severe shock, and adjusting to new circumstances was painful. One former organist and pianist appeared "very melancholy" and "not yet recovered from the shock received when her job was lost." Another former theater musician, who had a wife and three children, was "very despondent" and was "struggling to make ends meet without abandoning his profession." Among the displaced musicians who had found work in other fields, few reported their new jobs to be as satisfying or profitable as their old ones had been. A former organist who had made $42.50 a week now worked in a department store for $16. Another who had once made nearly $60 a week was now struggling to sell life insurance and had yet to earn significant income. Several musicians had moved in with their parents, in-laws, or children. Musicians with no jobs depended on charity. The Labor Bureau found that nine of the former musicians surveyed were "in dire need."[57]

THE PLIGHT OF musicians not only reflected the impact of the Great Depression but also spoke to the human consequences of technological change under capitalist control. The technology of early film production had helped musicians; silent films created thousands of new jobs for them. These jobs disappeared, however, when the talkies appeared. What technology gave, it eventually took away. Innovations in production processes did spawn new opportunities in Los Angeles and New York, where large orchestras recorded music for sound films. Yet these opportunities were meager compensation for the loss of vastly larger numbers of well-paying jobs.

For musicians, technological change eliminated whole categories of jobs almost overnight, with little regard to seniority or skill levels. Indeed, with one bold stroke and little warning the talkies obliterated a major segment of musical employment. Theater jobs had been the core of musical employment since the late nineteenth century and had functioned as a training ground for young talent as well. Unlike the new clerical workers in the business sector, theater musicians could not be retrained and given new tasks within the businesses that employed them. In fact, no one in the film industry made any attempt to help musicians. Their displacement, one industry leader noted, was a fact of technological progress.[58]

As their careers evaporated, thousands of instrumentalists who had believed—naively and wrongly, as it turned out—that live theater music was a fixture of public entertainment learned otherwise. Sound movies rudely forced musicians to realize that they were vulnerable to forces of technological change. As a result, by the early 1930s instrumentalists had lost the optimism they had had for three decades and more. "Unless you are possessed of the sort of push that comes in the back door after being kicked out of the front," a New York musician said, assessing the altered circumstances, "you will find nothing in this city of cutthroats." A few instrumentalists still hoped that orchestras would return to theaters, but the change was not, as Weber once suggested, a "temporary revolution." It was instead a whole new world.[59]

Three

Encountering Records and Radio

SOUND MOVIES WERE not the only challenge facing musicians in the interwar years. Records and radio also provided entrepreneurs new ways of reproducing and disseminating musical sounds and thereby displacing musicians. The story of the record and radio industries in the 1920s and 1930s shows again that the diffusion of new technologies had profound effects on the work environment and market power of musicians, and it clarifies, therefore, the impact of sound technology on labor relations. As music businesses provided ever-increasing audiences with less and less expensive entertainment through greater efficiency and economies of scale, more and more musicians worried about their jobs and careers and challenged those threats as best they could.

WHEN AMERICA entered World War I, four firms dominated the record industry. The Victor Company, whose flat shellac discs were acoustically much superior to the cylindrical recordings pioneered by Edison, was the industry leader. From 1915 to 1917 Victor accounted for perhaps half of all record production and sales, and in 1918 it produced nearly thirty million records. The Columbia Phonograph Company, the Brunswick Company, and Edison's National Phonograph Company accounted for most of the remainder, while a few dozen minor firms shared the rest of the market.[1]

Although many Americans spun these early recordings in handsome Victrola cabinets in their living rooms, most enjoyed them in coin-operated machines in commercial places. Early in the century several thousand of these machines, each consisting of a phonograph, hand crank, horn, storage battery, and mechanisms for accepting coins, were in operation. Despite the fact that they broke down frequently, the machines were highly profitable. In 1906 the Automatic Machine and Tool Company of Chicago, apparently the most successful manufacturer of coin-operated phonographs, introduced a successful cabinet model featuring a selector and record changer that gave listeners a choice of recordings. "Slot parlors," which featured up to eight or ten of these phonographs as well as other coin-operated devices, suddenly became important locales of public entertainment.[2]

Despite the sophisticated look of this forerunner of the jukebox, the phonograph provided only limited competition to coin-operated player pianos, to which it was acoustically inferior. At the turn of the century the Aeolian Organ and Music Company produced about seventy-five thousand player pianos a year; but it was the Wurlitzer Company of Cincinnati that, in 1898, introduced the coin-operated version. In 1910, after dozens of other firms had produced their own models, Wurlitzer advertised fifty different player pianos, at prices ranging from $1,500 to $10,000. Proprietors of cafés, bars, skating rinks, bowling alleys, and other such places found these "nickel-grabbers" profitable. Within six months of purchasing one of them, some proprietors reported a return on their investment of as much as 100 to 300 percent. In the early twentieth century, when a few familiar songs like "After the Ball" and "Maple Leaf Rag" could keep an audience happy, player pianos filled many commercial needs.[3]

Critics at the time believed that such devices trivialized the musical experience and degraded its moral and aesthetic value. Recorded music, they warned, would homogenize musical culture at the expense of distinction and art. Such critics were no doubt in part concerned that recordings were helping to bring forms of "disreputable" music into mainstream culture. In the early twentieth century, though many, perhaps most, consumers still favored traditional folk or classical music, growing numbers preferred the newer and less socially acceptable sounds of blues, jazz, and even rag. Columbia's 1923 recording of "Down-Hearted Blues" by singer Bessie Smith, one of the highest-paid African Americans in vaudeville, sold 780,000 copies. The recordings of jazz pianist Thomas "Fats" Waller, who established his reputation as an organist at the Lincoln Theater in Harlem, were also big hits, especially his "Honeysuckle Rose" and "Ain't Misbehaving." But the relationship between technology and culture these examples illustrate was an interactive one. The popularity of certain musical styles determined the kinds of music the industry produced and marketed.[4]

Recordings had important implications for musical education, and thus for musical culture. Most notably, they made aspiring musicians less dependent on traditional sources of training: family members, private teachers, and music schools. Ambitious youths could learn to play by listening to recordings over and over again. One cornetist, Jimmy McPartland, explained that he and his high school friends learned to play jazz that way. "We'd have to tune our instruments up to the record machine, to the pitch," McPartland recalled, "and go ahead with a few notes. Then stop! A few more bars of the record, each guy would pick out his notes and boom! We would go on and play it." That this was not a unique experience is suggested by the fact that some prominent musicians were reluctant to make recordings lest other instrumentalists use them to duplicate their style. In 1916, for example, Victor offered to record Freddie Keppard's Original Creole Band, only to be told, "We won't put our stuff on records for everybody to steal."[5]

Instrumentalists who did make records discovered at once that recording was quite different from other forms of performing. A recent collection of interviews with musicians has shed light on just how the new sound technology affected musical performance. Ernest L. Stevens, a pianist employed by Edison, told his interviewers that making records was stressful work: "If you were making a record, you had to be so careful." Sometimes, Stevens said, "I'd go all the way through, make a perfect record . . . and about a second before the end my finger would slip," and "the whole recording would be thrown out." Stevens's tenseness was a product of his boss's demand for precision. "I shook like a leaf for the first record," he recalled, "and I did the same thing for the six hundredth." Other musicians thought recording stifled creativity. Violinist Samuel Gardner, for example, who worked for both Edison and Victor, explained that performing close to a horn was difficult: "We had to stand sideways so that the sound went into the horn, and you had to avoid striking the horn." Playing under such conditions, he added, "was an awful battle."[6]

Earnings from recording work varied widely. Until local unions fixed wages in record companies, instrumentalists took whatever they could get. In 1911 Samuel Gardner earned $10 for each of the few days a week he worked for Edison. Although he was often at the studio ten hours a day, he spent much of his time waiting for the busy inventor to focus on music projects. At Victor several years later, Gardner received $35 for each selec-

Musicians at work at the Okeh Records recording studio, circa 1925. The primitive microphone, a large horn visible at the rear of the studio, called for forceful playing. Okeh turned out a series of jazz records that featured now-famous musicians and singers like Louis Armstrong and Bessie Smith. *(Bettmann Archive)*

tion he recorded. In contrast, recording stars, of whom the biggest was Enrico Caruso, the "Greatest Tenor of Modern Times," made fortunes. In 1904, when orchestral recordings were still rare, Caruso signed a contract with Victor that paid him $4,000 for ten "sides" of records.[7]

Musicians soon began to wonder whether recordings of popular artists and songs would undermine the demand for live music. For a time, however, recorded music was too scratchy to pose a serious threat, even though it played in commercial places and offered a few performers a way to supplement their income. Joseph N. Weber, president of the American Federation of Musicians, told AFM conventioners in 1926 that the phonograph had boosted public appreciation of music, and with it employment opportunities for musicians. "Instead of proving a development calamitous to our profession," he explained, "[the phonograph] has rather proven a boom." Recorded music reached the "smallest hamlets" of the country and in so doing advanced "the love of music among people" as well as "employment opportunities of musicians."[8] But the continued improvements in the quality of sound recordings concerned even Weber. Technicians at Western Electric had recently learned to convert sound waves into electrical impulses, which they then amplified and applied to the recording process. The advent of electrical recording, with its complex microphone and wire systems, produced recordings of unparalleled clarity and range. Manufacturers with heavy investments in traditional recording methods viewed the new technology with concern, but the fledgling radio industry adopted it at once. By the early 1920s, as a result, radio was offering sound reproduction of sufficient quality to attract rapidly growing audiences.[9]

FROM THE OUTSET, radio broadcasters used music to attract listeners. In fact, music was part of the first radio broadcast. When a former Edison employee, Reginald Fessenden, made his famous "continuous wave" transmission from Brant Rock, Massachusetts, on Christmas Eve in 1906, he featured a violin solo by himself. Other experimental broadcasters were soon transmitting performances of well-known vocalists and instrumentalists. But pioneers in radio did not rely solely on live performers; they also broadcast phonograph records and recordings of player pianos. Initially, then, live and recorded music complemented and competed with each other over the radio airwaves.[10]

The first commercial radio stations, however, used little recorded music. Because early recording technology depended on lung power, early phonograph records sounded tinny and amateurish in comparison with live performers. Radio audiences therefore much preferred live performances. In 1922, in apparent endorsement of that preference, the Commerce Department, to which the Radio Act of 1912 had given regulatory power over the airwaves, prohibited for-profit stations from broadcasting recorded music, including the music of player pianos. Justifying the prohibition, Secretary of Commerce Herbert Hoover reminded broadcasters that the airwaves belonged to the people and the public good demanded that radio stations use local talent and educational material. One of the responsibilities of publicly licensed broadcasters, Hoover maintained, was to create jobs in local communities. Broadcasters felt that they had no choice but to submit to Hoover's edict, for in the aftermath of the government takeover of the railroads during World War I, the fear of a takeover of radio as a public utility was understandably strong. As if to concentrate that fear, Congress had before it several proposals to nationalize radio.[11]

While all radio stations still shared the same wavelengths, as they did initially, local instrumentalists and musical groups performed on local programs across the nation. Radio audiences heard their performances interspersed with local weather reports, local and sometimes national sports scores, and children's programs. Schedules advertised in newspapers in 1924 helped audiences pick up live concerts. On Tuesday, February 3, for example, the *New York Times* advertised the weekend programs of more than forty stations, almost all of which included musical performances. On the following Sunday audiences on the East Coast might have heard the Meyer Davis Orchestra from WFI in Philadelphia or the Colonial Orchestra from WEAN in Providence. In the Midwest the available fare included music by the Rock Island Railway Orchestra from WOC in Davenport and the sounds of a "mixed quartet" from Zion, Illinois. Station KHI in Los Angeles scheduled four live concerts from eleven A.M. to three P.M. including one by the popular Coconut Grove Orchestra.[12]

The memoirs of singer Dorothy Stevens Humphreys, one of the first radio performers, illustrate the nature of early radio broadcasting. Humphreys began a long and successful career in radio in Columbus in 1920, when she was twenty-one years old. She later recalled her early experience:

> One day the phone rang and I was asked if I would like to be the first woman singer on the new stations? Would I?!!! But what would it be like? How did they do it? I wasn't the least bit afraid—just curious and excited. The day came! At WBAV—owned by the Erner and Hopkins Electric Company—I was ushered into a room, where the walls were completely covered with heavy burlap and near the piano was an odd looking contraption—a large wooden chopping bowl with something in the center (a microphone) on a metal stand. I was to stand in front of it and on signal—sing. . . . We did a half hour program and were quietly ushered out. It was over—but did anyone hear it? They did! The calls and mail [were] the answer to that—and my family just couldn't believe it—they had heard my voice coming over the air![13]

After this performance Humphreys sang for a second Columbus station, WCAH. "I sang in the [station owner's] living room," she recalled. "The music went over a telephone wire to the garage in the alley where the radio equipment was." The experience was a "glorious privilege," she said later, one that filled her with a sense of "great joy and fulfillment." Subse-

On the air at WJZ, New York, 1922. The first radio broadcasts sent the live performances of singers and pianists into the homes of thousands of set owners. At the
microphone stands soprano Luellea Melins. WJZ became the flagship station of the
National Broadcasting Company's Blue Network—later, the American Broadcasting
Company. *(Bettmann Archive)*

quently Humphreys worked for three other Columbus stations, and eventually she performed on eight programs a week. Her accompanists ranged
from a fourteen-piece orchestra to a string quartet, a musical duo, and occasionally a single pianist.[14]

Like Humphreys, other early performers found broadcasting novel and
exciting. In fact, when station owners refused to pay them, as they usually
did in the early years, musicians and singers worked for nothing. Broadcasters maintained that the publicity the artists received was compensation
enough for their services. Radio appearances did in fact boost the careers
of musicians, especially singers and bandleaders. Nevertheless, broadcasters soon found it necessary to pay performers. The first pressure to do so

came from the American Society of Composers, Authors, and Publishers (ASCAP), which songwriters and music publishers had formed in 1914. In the early 1920s, when royalties from the sale of records and sheet music were declining, ASCAP set out to make up the loss through radio, which was already the principal outside source of family entertainment. Accusing broadcasters of violating copyright laws, ASCAP threatened to sue stations that refused to pay royalties to the owners of the songs they aired, and it placed advertisements in trade journals portraying recalcitrant broadcasters as new exploiters of labor.[15] The pressure paid off when David Sarnoff, president of Radio Corporation of America (RCA), the leading manufacturer of radio sets, called radio "a niggardly buyer" of musical talent and suggested that the future of broadcasting depended upon musicians being paid. "Radio must pay its way," Sarnoff said in early 1924, "as [do] thousands of theaters, dance halls, and cabarets."[16]

AFM locals endorsed Sarnoff's view and began demanding wages for radio work. Kansas City musicians were among the first to notify broadcasters that union members would no longer perform without pay, and thanks to union control over musical services there, instrumentalists began earning $4 each per radio appearance. "Prefer[ring] pay to glory," musicians in Chicago also refused to play without pay. "Radio broadcasting," they proclaimed, "has developed to such an extent in the past year as to become a serious menace to the professional musician." The Chicago union demanded that broadcasters hire orchestras on the basis of three-hour periods. As a result, in late 1924 Chicago musicians were receiving $8 each per radio engagement, regardless of whether they played three full hours.[17] James C. Petrillo, president of the Chicago local, explained the reasoning behind the union demand: "We are forced to this action by the falling off in demand for orchestra and band music since broadcasting has become popular." The public, Petrillo complained, instead of going to live performances, was staying at home and listening to radio. "People sit back in their homes and enjoy our performance," he reasoned. "Parties enjoy dancing to the faraway invisible orchestra. This is all right, but if it brings unemployment to our ranks we are justified in levying a moderate fee for our protection."[18]

To secure their position in radio, musicians' unions began demanding that stations hire orchestras of their own, sized according to station wattage, which largely determined audience size. The most powerful stations were asked to employ bands of at least twenty-three pieces, while the smallest stations could hire as few as two musicians. To enforce the demand, locals prohibited their members from performing for uncooperative broadcasters. The union monopoly on musical services and the dependence of radio on live performances ensured compliance with the demand.[19] This system guaranteed musicians steady work in radio. Buoyed by it, bandleaders began signing yearlong contracts with broadcasters, and the resulting radio orchestras earned handsome wages working twenty to thirty-five hours a week, forty to fifty weeks a year. At a time when the average wage earner in manufacturing made less than $125 a month, radio musicians were earning as much as $250. By 1925 more than five hundred radio stations across the country were providing staff musicians better-than-average wages as well as playing time to improve and to advertise their skills.[20]

The popularity of staff orchestras increased as radio adopted "toll" broadcasting, a format by which broadcasters passed some of their programming costs to advertisers. Station WEAF in New York, owned by American Telephone and Telegraph (AT&T), introduced this format in 1922, when it offered commercial advertisers airtime at the rate of $100 per ten minutes. The format quickly evolved into commercial sponsorship of half-hour musical concerts by WEAF musicians, and broadcasters around the nation were soon following the example. On WJZ in New York, to illustrate the pattern, the Wanamaker Organ Company sponsored weekly organ concerts, while in Dallas the Magnolia Petroleum Company did likewise for a program of live dance music on WFAA. By 1925 more than 160 of the nation's 561 radio stations had commercial sponsors for at least some of their programming. Musicians thus benefited as advertisers came to rely on radio orchestras to generate consumer demand for their products.[21]

As business sponsorship increased, improvements in radio technology accelerated the rising popularity of the broadcast medium. Better amplification of loudspeakers improved the quality of sound and eliminated the need for audiences to wear cumbersome headphones, and soon receivers could be plugged into ordinary electrical circuits, thus eliminating the need for batteries. The dramatic drop in the price of radios further boosted the medium. In 1924 a radio cost more than $200, but only three years later Sears, Roebuck was selling Silvertone models for $34.95, and nearly ten million Americans owned radios. These developments increased not only radio sales but the number of radio stations as well. As long as stations broadcast meaningful amounts of live music, this meant additional jobs for musicians.[22]

What all of this suggests is that throughout the artisanal stage of the radio industry, musicians had great influence over broadcasters and broadcasting policy. The poor quality of recordings made live music indispensable to broadcasters, and broadcasters themselves were unorganized and in no position to resist musicians' demands. Under these market conditions, unions shaped hiring patterns, job classifications, wage scales, and working conditions in the industry. Two union demands were especially effective in producing and sustaining this outcome: minimum wage scales and minimum orchestra sizes.

CHANGING CIRCUMSTANCES soon threatened this outcome. Especially troubling for instrumentalists were the appearance and sudden proliferation of remote-control broadcasting. Broadcasters made "remotes" by running wires to hotels, theaters, dance halls, and nightclubs where transmitters picked up live musical programs and broadcast them over radio. This practice gave broadcasters access to new pools of talent, but it also reduced the need for staff orchestras. Places featuring live music gladly permitted remote broadcasting of the programs, for it meant free advertising, often on a national scale, and musicians in search of exposure welcomed the opportunity to get their music broadcast at no charge to themselves. To protect staff orchestras, local unions had to control remote broadcasts. Thus, in Chicago the union began permitting remotes only on stations employing staff orchestras. In many smaller cities, however, that option was not available to union locals.[23]

A more serious threat to radio musicians arose when stations in different localities began broadcasting programs simultaneously. In the early 1920s broadcasters discovered that news, sporting events, and musical programs emanating from one station could be carried by telephone lines to other stations, and thereby be transmitted to audiences far from the site of the broadcast. WEAF in New York and WNAC in Boston first demonstrated this technique, in 1923, when they simultaneously broadcast a five-minute saxophone solo. Soon thereafter a chain of twelve stations linked themselves together to broadcast one of the nation's first commercially sponsored shows, the *Eveready Hour*. In 1925 a chain of twenty stations simultaneously aired a speech by President Calvin Coolidge.[24] These developments enabled Americans to hear of national events as they happened, but they also threatened the livelihoods of many musicians. Events in Boston in 1924 illuminated the problem. There, three thousand people had bought tickets for a concert featuring the renowned violinist Frits Kreisler when a local radio station announced it would broadcast the concert live.

As a result of the announcement, more than half the ticket holders asked for refunds.[25]

At the 1926 convention of the AFM, president Weber addressed the growing concern over the impact of radio on employment. Radio, he predicted with typical optimism, "will broaden the knowledge and desire of the people for music, hence will ultimately increase the employment of musicians." Asked whether broadcast music might not replace live music, Weber responded in the same vein. "Radio services cannot always be had when wanted," he said, and would in any case be impractical "where the rendering of a set program is necessary at a certain time." He assured the convention that "no transmitted musical service will everlastingly displace the desire of the public for personal services rendered by the artist in the presence of the public."[26]

The rapid rise of radio had already cast doubt on Weber's assessment. Even as he spoke, the sudden appearance of a new national network linking radio stations from coast to coast was overshadowing all previous developments in the industry. This revolutionary innovation was a product of aggressive new strategies of both integration and diversification on the part of RCA. In 1926 RCA purchased the powerful New York station WEAF and created a subsidiary, the National Broadcasting Company (NBC), to transmit the station's commercial programming to broadcasters that affiliated with NBC. The toll-broadcasting format dictated the economics of the programs thus broadcast; NBC, however, paid affiliated stations between $30 and $50 for each commercial program they broadcast. By linking stations across the country, this arrangement created the largest, most effective advertising medium the world had ever known. In 1927, advertisers paid NBC about $7 million for its services.[27]

RCA followed up this success with a second network built around another powerful New York station, WJZ. By creating the second system, called the Blue Network because of the color of the markers that charted its affiliated stations on a map, RCA hoped to capture both ends of the music market. The Blue Network featured nonsponsored programs that appealed to middlebrow and sometimes highbrow listeners who preferred cultural or educational programs having little popular—or commercial— appeal. The network carried such things as concerts of symphonic music performed by staff orchestras, which NBC funded but whose costs it partially covered by charging affiliated stations $50 to $90 an hour for broadcast rights to them. Some stations received programming from the original Red Network as well as the newer Blue Network.[28]

Independent stations immediately recognized the value of network affiliation. The networks provided them with vastly improved programming, which in turn increased the size of their audiences and encouraged businesses to market their products on radio. Advertisers willingly paid higher fees for the chance to place advertisements between network shows. Network programming also gave stations time to develop programs of their own, programs that occasionally attracted national sponsors and thus added revenue and prestige. But in doing these things the networks notably reduced the need for local performers, including musicians. In 1930 NBC had more than seventy affiliates and net profits in excess of $2 million.[29] The future for most radio performers obviously lay with the networks.

The success of NBC prompted the creation of two competing networks. The more important of the two emerged in 1927, when William S. Paley of Philadelphia purchased two regional broadcast alliances and merged them into the Columbia Broadcasting System (CBS), with WABC in New York as its flagship station. The immediate success of CBS resulted largely from Paley's innovativeness. With no income from the manufacture of radio sets to offset his broadcasting costs, Paley adopted aggressive strategies to attract advertisers while he revolutionized the relationship between the network and its affiliates. CBS provided nonsponsored programs to affiliates free of charge in exchange for the right to preempt their "prime" airtime in the early-evening hours. Once a show obtained a sponsor, CBS paid affiliates to carry it. One advantage of these formulae was that they immediately guaranteed advertisers specified time slots, which brought greater predictability to marketing. As for program content, CBS consciously adopted a popular, lowbrow approach. It presented the music of popular bandleaders like Paul Whiteman and equally popular comedy acts and soap operas. As a result, CBS quickly became the second largest network, with nearly eighty affiliates in 1931.[30]

The third major network appeared in 1934, when several large independent stations pooled their resources to form the Mutual Broadcasting System (MBS). Unlike NBC and CBS, Mutual was not itself a corporation but a federation of independent stations that exchanged programs. Each station agreed to create its own programs, secure sponsors for them, and share the programs with other federated stations. Because NBC and CBS had already signed the largest and most profitable stations, Mutual looked for affiliates among small stations, including those in small cities outside the populated areas of the East Coast. Some Mutual "quasi affiliates" also affiliated with NBC or CBS. In the late 1930s, when more than a hundred stations were regularly broadcasting Mutual programs, the network's key stations were WOR in Newark, New Jersey, WGN in Chicago, WLW in Cincinnati, WXYZ in Detroit, and KHJ in Los Angeles.[31]

The rise of MBS completed an industrial structure that remained undisturbed until the early 1940s, when federal regulators forced NBC to divest itself of the Blue Network, which became the American Broadcasting Company (ABC). At the center of this structure were the three major networks, which owned and operated a handful of key stations in major cities. Leasing telephone lines and transformers from AT&T, the networks sent their programs to some three hundred affiliates, which used the programs to supplement their own locally produced shows. On the periphery of the system were roughly an equal number of low-powered unaffiliated stations. Some of the latter broadcast from rural areas and catered to the concerns of farmers or regional audiences, while others had ties to universities or civic organizations and operated on a part-time basis. Most of them relied on news reports and recorded music and seldom, if ever, featured performances by local musicians. Many were outside the jurisdiction of musicians' unions.[32]

By the early 1930s, then, radio had moved beyond its artisanal stage. It was no longer an industry of fledgling stations struggling to keep abreast of technological innovation. It had instead become an oligopolistic industry dominated by a small number of large, integrated firms competing in a national market. These firms were monuments to the efficacy of financial and bureaucratic efficiency. In 1935 the central administration of NBC included departments of management, engineering, programming, sales, and traffic subdivided into eastern, central, and western divisions. There were also special departments, including promotions, station relations, and public relations. The key stations in the NBC network were themselves multidivisional firms. The money that flowed to network headquarters in New York best exemplified the magnitude of the firm's operations. In the midst of the Great Depression, 1935, NBC had eighty-seven affiliates, gross receipts of $26.7 million, and net profits of $3.6 million. It was only one of the subsidiaries of RCA.[33]

The triumph of network broadcasting took place against a background of economic hardship and social despair. Beginning in the early 1930s, the industrialized world was in the throes of a deep and persistent depression. From 1929 to 1932, stock prices on the New York stock exchange fell by approximately 75 percent, while the value of goods manufactured dropped 50

Crooner Rudy Vallee and his Connecticut Yankees performing in New York in 1932 for one of NBC's most popular commercially sponsored shows, the *Fleischmann's Yeast Hour.* Standing left to right are Chic Johnson, Ole Olsen, and the ever-popular Vallee himself. *(Ole Olsen Collection, Cinema–Television Library, University of Southern California)*

percent and several thousand banks and countless numbers of other businesses failed. The ranks of the unemployed swelled accordingly, from 3.2 percent of the nonfarm workforce in 1929 to more than 25 percent by 1933. In many jobs wages fell so low that workers who continued to work had difficulty surviving. In such a context, the economic performance of the radio industry was all the more remarkable.

TALENTED MUSICIANS in media centers were among those who prospered in the new corporate setting in radio. In New York, Chicago, Los Angeles, Detroit, Cleveland, and a few other cities where network programs originated, musicians found expanded opportunities. *Variety* testified to the changing times: "While musicians and musical organizations in general are fighting the mechanical age in the music biz, a select few are praying for it to stick. These are the radio musicians." Some instrumentalists in New York were earning up to $800 a week, playing for so many shows that cabs had to rush them from one studio to another. "I saw it!" recalled the wife of one musician. "They'd run from NBC and change their jackets on the way. They had guys who'd carry their horns and everything and get them on to the next show. This went on and on. They were making money like millionaires."[34] By 1935, when network broadcasting was in full bloom, more than one thousand musicians were making their living in radio. In addition, advertising agencies and program sponsors provided single-engagement work for many others. For most "commercial" programs, the networks and advertisers secured the services of well-known bandleaders and orchestras. In fact, NBC organized the Artists Bureau and Concert Service to ensure the services of star performers for its programs.[35]

George Olsen was one of the first bandleaders to realize the potential of network programming. Having established a reputation in vaudeville in New York, Olsen had the honor of conducting his orchestra on NBC's inaugural broadcast, on November 11, 1926. In the aftermath, Olsen was hired at a salary of $2,500 a week to conduct his orchestra on a weekly program on NBC sponsored by Canada Dry. The Olsen orchestra, which featured his wife, Ethel Shutta, as vocalist, thus became familiar to millions of Americans. Olsen opened the performances with "Beyond the Blue Horizon" and closed them with "Going Home Blues." The popularity of his NBC program helped Olsen secure work outside radio for years thereafter, in some of the nation's most popular theaters, hotels, and dance clubs.[36]

Network radio popularized many bandleaders and orchestras by remote broadcasting. In the mid-1930s remote broadcasts of bands like those of Fred Waring, Guy Lombardo, and Ben Bernie were among the most popular programs on radio. NBC and CBS carried remotes nightly at that time, from eleven P.M. to one A.M.. while Mutual did so from eleven to two. Bandleaders featured on these broadcasts typically translated the free publicity and exposure into more jobs at higher wages. So valuable did remote broadcasts become for the bands and their leaders that some of them accepted lower wages in order to play in places with remote outlets.[37]

Remote broadcasts helped give rise to the swing era, a distinct and exciting time in the history of American musical culture. Throughout the 1930s and 1940s, and especially between 1935 and 1945, several dozen nationally known "big bands" of ten to twenty pieces traveled across the country performing for a night or a week or two in hotels, dance clubs, and other such places. The typical bands played swing, a blend of rapid rhythmic beats and New Orleans–style jazz whose melodic sophistication preserved the improvisation that made jazz so popular with artists and au-

The Duke Ellington Orchestra recording in New York, 1937. Recording technology continued to improve during the 1930s, greatly expanding the audience for and the popularity of dance bands. *(Courtesy of Ray Avery)*

diences alike. The melodic riffs of horn sections were especially distinctive in much of this music. Benny Goodman's rendition of "Stompin' at the Savoy," Duke Ellington's "Take the 'A' Train," and Glenn Miller's "In the Mood" are classic examples of swing.

Traveling with the big bands was often exciting, but working conditions were frequently anything but glamorous. The bands sometimes played six or seven one-nighters a week, each in a different city. At the most hectic pace, musicians barely had time to shave before performances or to eat afterward. They often returned to the bus after a three- or four-hour performance only to have to travel through the night to the next job. On the road, musicians complained of bad food, lack of sleep, and a host of other inconveniences. Frank Sinatra, who traveled with the Harry James Orchestra for a while, recalled the routine: "There's nothing to beat those one-nighter tours, when you rotate between five places around the clock— the bus, your hotel room, the greasy-spoon restaurant, the dressing room (if any) and the bandstand. Then back on the bus to the next night's gig, maybe four hundred miles away or more."[38] But it was tours like this that

Remote broadcast of the Nat "King" Cole Trio, Hollywood, 1941. During the 1930s and 1940s radio stations relied on live music to attract audiences, especially during the late-night hours. KFWB picked up this performance of the soon-to-become-legendary Cole playing piano at Music City on the corner of Hollywood and Vine. Wesley Prince on bass and Oscar Moore on guitar accompanied Cole. Standing in the back of the room, at the upper left of the photograph, was Glen Wallich, founder of the nightspot—and of Capitol Records. *(Courtesy of Jerry Anker)*

made Sinatra, as well as other crooners like Rudy Vallee and Bing Crosby, a national star, first on radio and then in the movies.

Vocalists like Sinatra occupied a pivotal spot in the musicians' world. Indeed, they defined the era in terms of popular music. Popular singers added personality and even sex appeal to musical performances, and thus enhanced a band's value in the leisure market. During road trips vocalists often took on added responsibilities, such as caring for the music library or making travel arrangements. One reason for these additional chores was that singers had no union, and as a result they had less control over the conditions of their work and the wages they received. In fact, many of them worked for lower wages than the sidemen who accompanied them.

Vocalists learned to expect and deal with such problems as out-of-tune pianos, overly loud bands, poorly keyed arrangements, and inferior public-address systems. Big bands often featured female vocalists, some of whom—Doris Day, Peggy Lee, Dale Evans, Lena Horne, Ella Fitzgerald, and Billie Holiday—became stars in their own right. In a predominantly male business, female vocalists faced their own challenges, among which loneliness, male chauvinism, and the difficulty of ironing a dress in a moving bus were characteristic. Doris Day, who sang for the Barney Rapp and Les Brown bands, recalled the traveling experience from a woman's point of view. "Being on the road is not easy," she said, "especially for one girl among a lot of guys. There's no crying at night and missing mama and running home." Day nevertheless believed that her experience on the road had made her a "stronger," more "disciplined" performer. Like Day, many vocalists who began singing during the swing era remained stars long after that era passed.[39]

Big bands made up of black musicians confronted the challenge of traveling in segregated regions of the country where racism was blatant, even dangerous. "In those days you went South at the risk of your life," one black instrumentalist recalled. Perhaps the most popular of these traveling bands was led by the charismatic Cab Calloway, one of the few African Americans who appeared regularly on network radio. When Calloway's New York–based group toured Texas in 1935, it found itself in several perilous situations, especially at dances with mixed audiences. Bassist Milt Hinton toured with the band and recalled the experience: "At those dances the prejudice was terrible. Some [whites] would say, 'I'd pay a three hundred dollar fine just to hit one of those boys.'" On more than one occasion, Hinton said, promoters had to take band members to a private room, "to keep the people from getting at us." Racism was equally pronounced in parts of the Far West. "Going west was the same as being in Georgia or Mississippi," reedman Garvin Bushell said. "You couldn't stay there, couldn't eat there, couldn't go out."[40] Indeed, many experiences that white musicians found manageable, like getting food and hotel accommodations, black musicians found problematic.

STRUCTURAL CHANGES in radio presented musicians outside media centers with especially serious challenges. Network programming meant that superior players in a few large cities provided better music for listeners across the country than local bands could provide for those in their communities. Musicians in St. Paul (Minnesota), St. Petersburg (Florida), and other places across the country considered network broadcasts a form of unfair competition. Local stations, they complained, used network programs to avoid hiring local musicians, and thus to eliminate local orchestras. At the AFM convention in 1932, delegates from Local 77 in Philadelphia warned that musicians must control network broadcasts or else the "hook-up process" would reduce radio work for musicians to that of "a few highly trained specialists in two or three of our large cities."[41]

Other structural changes in the industry compounded the problems of musicians outside national media centers. The Radio Act of 1927 created a presidentially appointed Federal Radio Commission (FRC) to regulate use of the airwaves. Although unions, among others, could challenge the commission's decisions in courts, in fact they rarely did so, in part because the commission had the major voice in deciding who received new and renewed broadcast licenses. Unfortunately for musicians, the commission let lapse the policy forbidding commercial stations to use recorded music in programming. The commission agreed that radio had a responsibility to hire and foster local talent, and it even secured pledges from veteran as well as would-be broadcasters that they would not fill the airwaves with recorded music. But in practice the commission did not question programming policies. In fact, most commissioners accepted the broadcasters' own argument that local talent could not meet the standards of quality their audiences demanded. "It is true," the FRC stated, "that in the smaller communities which do not have adequate original program resources, the use of phonograph records may fill a need." The commission asked only that "mechanical reproductions" be identified as such. When the Communications Act of 1934 transferred regulatory power to the new Federal Communications Commission (FCC), the new agency continued to issue and renew licenses with little regard to local programming policies.[42]

For a time, trends in the record industry softened the impact of these developments. Although the industry had shown new signs of life in the mid-1920s when electricity and microphones made their way into the recording process, the Great Depression reversed the industry's growth. Between 1929 and 1933 the number of records sold annually fell from approximately 100 million to 10 million, and in the latter year recording companies manufactured only 2.5 million records.

But as sales and production plummeted, new developments transformed the business of recording and speeded its recovery. In 1929 RCA bought control of the financially troubled Victor Company and brought more sophisticated technologies as well as new management skills to the record industry. Over the next few years RCA-Victor introduced and popularized electrical recording as well as slower-turning, longer-playing, and more flexible discs. The company also fostered the development of efficient coin-operated music machines. Holding twelve to twenty-four records that distributors could easily—and therefore frequently—change, the jukebox quickly superseded the player piano as the nation's favorite source of cheap musical entertainment. In 1935 six companies produced a total of 120,000 of these machines. Jukebox operators quickly became the best customers of record manufacturers, purchasing roughly a third of all records sold in the mid-1930s. The formation in 1934 of the Decca Company, which specialized in the production of inexpensive (35-cent) records, coincided with and contributed to the rise of the jukebox business.[43]

Decca's inexpensive records played a key role in the popularity of swing music, which soon became a boon to all facets of the record industry, but especially to the jukebox business. Across the country thousands of hepcats and bobbysoxers pumped nickels into jukeboxes and jitterbugged to recorded swing music. By 1935 the craze had induced the industry to introduce long-playing records.[44] Again, cultural and technological changes coincided and reinforced each another. Technical innovations in recording encouraged new musical styles and dance steps, which in turn reshaped production technology.

By popularizing new recordings and musical styles, the jukebox further enhanced the star value of bandleaders and orchestras. Tommy Dorsey, whose up-tempo band was popular throughout this era, acknowledged the importance of the jukebox in his rise to stardom. "It's the sale of the records that makes a band," Dorsey explained, "and it's the jukeboxes that use the most records. These kids come out for lunch or recess and they pack into these soda parlors feeding nickels into the jukeboxes." The music machines, Dorsey recognized, drew large and enthusiastic audiences to his and his band's performances. To other musicians, however, the jukebox was a threat. In bars, dance halls, ocean-liner saloons, and countless other places where instrumentalists had once worked, coin-operated music machines became profitable alternatives to live performers. The precise step-by-step effect of jukeboxes on musical employment has never been documented definitively, but by the early 1940s AFM leaders believed that the machines had already cost musicians at least eight thousand jobs.[45]

The development of electrical transcription (ET) represented another technological advance with problematic consequences for working musicians. ETs were slow-spinning sixteen-inch discs sold only to radio stations. The Vitaphone Company's Harold J. Smith, who invented these discs, designed them to accompany silent movies, but they soon became more valuable to radio stations than to movie houses. With fifteen minutes of programming on each side, the discs gave broadcasters complete musical concerts to substitute for live local performances and even for network programs. The popularity of transcriptions, however, lay in the advantages they offered advertisers. Working with advertising agencies, transcription manufacturers made the big discs with prerecorded advertisements, or left blank spaces on them so that broadcasters could insert advertisements of their own.

Transcriptions thus provided a low-cost, highly efficient vehicle for advertising, and one independent of staff orchestras and the networks. The flexibility they thus afforded for the design of marketing strategies enabled advertisers to design specific programs for specific audiences with great precision and effect. In 1932, only two years after the introduction of transcriptions, approximately 75 percent of all stations were using them for some part of their programming needs, and advertisers were spending roughly $10 million to broadcast them. As markets for consumer goods expanded in the 1920s and contracted in the 1930s, everyone from car manufacturers and cosmetics makers to oil companies and local merchants sponsored transcribed programs to boost sales. Growing numbers of radio stations contracted with leading transcription companies like World Broadcasting Service and Standard Radio Library for fixed numbers of discs on a monthly or bimonthly basis.[46]

Transcriptions posed challenges to networks as well as staff orchestras. Although transcriptions solved problems related to program scheduling and time differences across the country, they also reduced the dependence of local broadcasters on network programming and in doing so threatened network profits. At first the networks discouraged the use of transcriptions, refused in fact to air them, and promoted the superior quality of their own live programs. After a while, however, they entered the transcription business. In 1934 RCA-NBC created a subsidiary, RCA-Thesaurus, to manufacture and sell or rent transcriptions. Initially this "transcription library service" distributed only NBC transcriptions to affiliated stations, but it soon began selling discs to other stations as well.[47] In these and other ways musical employers as well as musical workers adjusted to new technologies to survive in a rapidly changing industrial environment.

Big bands thrived in the new environment. During the depression decade transcription firms recorded the performances of the bands of

Benny Goodman, Woody Herman, Ray Noble, and other favorites for later radio broadcasts. Some bandleaders worked for several transcribers simultaneously and pocketed small fortunes, while sidemen had to satisfy themselves with union scale for transcribed performances—in the 1930s, about $24 for a three-hour recording session. But transcription also created problems. At least a few small companies pirated transcribed programs "off the lines" as they were broadcast, and though the pirated recordings were of poor quality, the companies sold them cheap to radio stations. In 1934 AFM president Joseph Weber complained that some broadcasters themselves were recording live radio performances for subsequent illegal sale. "Some stations," he said, "steal the music and sell it, and the musician is absolutely unprotected against such piracy." This practice, known as bootlegging, was far less a threat to musicians than was the production of legitimate transcriptions, which were a viable alternative to live radio performances. Transcription notably reduced the need for staff orchestras; even network musicians worried that it would eventually eliminate live radio work.[48]

INCREASING COOPERATION between broadcasters over labor-related issues was another threat to musicians. Since the late nineteenth century, in industries that became capital-intensive and structurally integrated, employers had pooled their resources to neutralize the potentially ruinous threats of competition and labor strife. The effort achieved new levels of cooperation around the turn of the century, when employers formed permanent trade associations dedicated to, among other things, resisting trade unionism. Builders in Chicago, for example, formed the Chicago Contractors' Council at this time to force unions in the building trades to abolish work rules that limited production and controlled the hiring process. The endeavor provoked a lengthy strike during which the council maintained its unity and helped deal a serious blow to Chicago trade unions. In the 1920s employer associations became far larger in scope and more aggressive in their antiunionism. Led by the National Association of Manufacturers (NAM), they lobbied lawmakers for favorable labor laws and launched public relations campaigns to gain public support for their stance during labor conflicts. In pursuing these goals the associations did more than eliminate union work rules and hold down wages during the 1920s and 1930s; they also convinced millions of Americans that trade unions and trade union leaders were "un-American" because they interfered with a person's right to sell his or her labor in the marketplace. NAM president

John E. Edgerton stated the attitude of most employer associations when he denounced trade union leaders as "pirates who parade in the guise of workingmen's friends."[49]

Like many employer organizations, the National Association of Broadcasters (NAB) began in opposition to labor activism. Its origins went back to 1923, when a number of station owners organized to prevent ASCAP from "extorting" license fees for the use of music composed by ASCAP members. "In our indignation [over ASCAP demands]," a Chicago broadcaster said, "we . . . decided we would form an organization." This pattern of one organized interest group spawning an oppositional interest group has a long history. As John Kenneth Galbraith explained years ago, "Power on one side of a market creates both the need for, and the prospect of reward to, the exercise of countervailing power from the other side."[50] Broadcasters did not create the NAB simply to countervail the economic power of their labor force; they organized it, instead, to dominate the labor force and thereby maximize their margins of profit.

Size was but one measure of the NAB's power. Through a well-oiled public relations operation, the association maintained efficacious alliances with molders of public opinion across the country. Through their control of news programs and in other ways, broadcasters interpreted industrial developments to the general public, and their interpretations reappeared in the print media, and in turn in the collective perceptions of varieties of public organizations—religious circles, women's organizations, educational groups, and other reflectors as well as molders of public opinion. NAB spokesmen often discussed radio as if it were a public resource the rise of which was a consequence of the functioning of natural forces of science and technology. One of them even called radio the "tool of democracy." The effect of such depictions was to neutralize critics of the broadcasting industry by making them appear to be backward-looking opponents of social or scientific progress, or of the people's right to know and hear. The prestige of broadcasters and NAB officials as well as their clout in national advertising and their ties to national business interests all enhanced the influence of radio at all levels of politics. Broadcasters were thus able to influence legislation that, among other things, prohibited their employees' unions from trying, in collective bargaining, to curtail the use of recorded music in radio broadcasting.[51]

During the 1920s the NAB had been relatively weak compared with the AFM in collective bargaining. But by the early 1930s the NAB had become the collective voice of several hundred broadcasters, and the balance of power between the two organizations shifted noticeably. When that occurred the NAB became a significant force in the world of working musicians. Before 1930, union musicians faced little organized resistance from employers. Theater owners had sometimes banded together to resist actors or musicians, but their combined power before the advent of sound movies was never impressive, and unions had effectively protected the wages and working conditions of theater musicians. The rise of NAB, however, put musicians in a position not unlike that of other skilled workers who faced powerful employer associations bent upon obstructing union goals. The nature of the radio industry, a closely knit web of a few vertically integrated firms, not only centralized but also facilitated the exercise of employer power.

The NAB, moreover, was not a faceless hierarchy. As Weber and other union officials understood, NAB leaders were personalities whom the public generally perceived as benefactors but whom they had to deal with as aggressive enemies of labor. Among the leaders was William S. Paley, a calculating, assertive businessman out to maximize profits at the expense of anyone who challenged his purpose. Paley's father, a Chicago cigar manufacturer, had been among the first in that industry to introduce machinery that simultaneously sped up production and reduced skill levels. When his father moved the business to Philadelphia in 1919, young Paley became a factory manager and made his first mark on labor history. In his father's absence he staffed the factory with young, semiskilled women to tend the new machinery. Shortly thereafter, skilled cigarmakers across the city went on strike to protect their wages and working conditions, and young Paley used his wealth and charm to persuade the women to stay on the job. He apparently provided them refreshment parties, boat rides on the Delaware River, and escort services to and from work. Recalling this experience in labor relations, Paley later remarked, "I became conscious of the fact that my boyhood had ended and that there were things in the world I could do and do well."[52]

To AFM officers in the early 1930s, the music industry was a world turned upside down. The mass dissemination of music via records and radio produced not a musicians' bonanza but their newest and greatest problem. "The mechanical developments in radio have been so rapid," union leaders concluded in 1933, "that it has been a constant battle to secure for our membership even the smallest percentage of what should constitute their fair share of the profits of industry. Records, electrical transcriptions, remote control and chain hook-ups," they continued, "have all

contributed toward the complete elimination of the musician or the caus-
ing of each man employed to replace hundreds of men, just as in the case
of the sound picture. . . . The possibilities of destruction of employment in
all industries where sound reproduction is involved [are] simply appalling,
a single station may one day service the entire country."[53]

AFM OFFICERS responded to the challenge in multiple ways. First, they
turned to the federal government, warning the Federal Radio Commission
in 1929 that "the invasion of the radio field by canned music [is] destroying the advancement of art at its base by depriving musicians of the necessary means of livelihood." Just as they did in the ongoing battle against
sound movies, union officials argued that eliminating the sources of musicians' income hurt not only musicians but American culture itself. Even
gifted instrumentalists, they pointed out, find themselves victims of technological unemployment. Despite the substance of this appeal, the commission responded that it lacked authority to interfere with broadcasters'
use of recordings, and the appeal was in vain.[54]

Musicians reacted with understandable anger. At the AFM convention
in 1933, A. C. Hayden, president of the Washington local, denounced
broadcasters and federal officials alike. "They use their power against us,"
Hayden said of both groups, "whenever it is to their advantage and profit
to do so." Hayden likened radio's access to political influence to the power
the railroad industry once accrued by giving train passes to politicians. He
accused broadcasters of similarly bribing politicians, up to and including
the president: "In insidious ways, [broadcasters] ingratiate themselves into
the favor of government officials . . . [and] give the President and cabinet
officers [all] the [free campaign advertising] time they want, whenever they
want it."[55]

While Hayden castigated broadcasters and politicians, other convention
speakers raised more fundamental questions. How could the union help
the growing number of locals concerned about decreasing job opportunities outside media centers without jeopardizing the members of locals
in those centers whose work was partly responsible for the decrease? How
was infighting between musicians in large and small cities to be avoided?
What constituted a "fair share" of profits in the rapidly changing radio industry? As delegates debated these questions, the continuing loss of theater
work plus the general insecurities of the depression era compounded their
problems.

At the 1933 and 1934 AFM conventions, delegates offered various answers to the questions above. Those from Philadelphia recommended tariffs of $3 or $5 on each network program a station carried. Stations with staff orchestras would pay the lower tariff, those without orchestras the higher one. Several small locals suggested a boycott of networks that sold programs to affiliates without staff orchestras. Other delegates desperately urged the president of the federation "to immediately negotiate an arrangement [with broadcasters] in any manner that he sees fit to relieve the situation."[56] The most drastic proposal came from Chicago, where musicians had recently waged a bitter struggle with radio. After an unsuccessful attempt in 1931 to curb the use of recordings, James C. Petrillo of Local 10 proposed that the union confront the source of the recordings: record manufacturers. Specifically, Petrillo wanted manufacturers to place restrictions on the uses of their products, and he was ready to strike all record companies to accomplish that goal. His plan appealed to officers of small locals that had little to gain from record manufacturing, but the executive board rejected it. The board feared that such a strike would constitute a secondary boycott, an action designed to pressure a third party to force employers to comply with union demands. Such boycotts had been illegal since 1908, when a hat manufacturer in Danbury, Connecticut, convinced a federal court that a union-sponsored boycott of his products was a "conspiracy in restraint of trade," and thus a violation of the Sherman Antitrust Act. This ruling, which reflected growing juristic antagonism to organized labor, cost 197 members of the United Hatters Union fines of $240,000 and was bitterly resented by union leaders. Despite the implications of the ruling, Petrillo and some other AFM leaders believed that the courts would permit the proposed boycott because musicians worked in the recording industry.[57]

But Weber eschewed the kind of confrontation Petrillo supported, opting instead for "patient efforts" that he hoped would pay off in the long run. But in view of the sharp decline in union membership in the early 1930s, this kind of moderation looked less and less promising. Layoffs in theaters mounted, opportunities for work outside music evaporated, and musicians found it difficult even to pay their union dues. By 1934 the union was in full retreat. In five years membership had fallen from more than 150,000 to 100,000, costing the federation about $60,000 in annual revenue. These figures, plus Weber's belief in the futility of resisting technological innovations, do much to explain the union's reluctance to respond aggressively to the desperate circumstances of most of its members. "Workers set their faces forever in vain against the development of machinery," Weber had told musicians in 1929.[58] A few years later he told them again that nothing positive could come from pulling instrumentalists from the record industry: "The withdrawing of a handful of workers in an effort to hinder the continuation of an industry which represents an investment of billions of dollars would prove nothing else except that our conventions were devoid of proper discernment and our leaders were mere mental jugglers in their efforts to constructively meet changed conditions."[59]

Weber also feared that the action Petrillo proposed would endanger the opportunities new technology had already created and would continue to create in the future. Worried about the decline in union membership, Weber warned advocates of confrontation that "attempts to create employment in one direction may destroy other employment." The federation, he insisted, could not prevent broadcasters from importing canned music from other countries. "Musicians in London," he explained, "could in a short time litter this country with hundreds of thousands of records."[60]

At a critical moment in the history of musicians, then, their union decided to accept rather than resist technological change. Instead of an all-out strike to challenge the consequences of new technologies, Weber focused on encouraging the well-intentioned if impractical policy of the National Recovery Administration (NRA), a New Deal agency, to rejuvenate the economy through a "spread the work" plan. At the request of the NRA administrator in charge of the amusement trades, Weber agreed to a plan to aid out-of-work musicians by rotating the work available in motion pictures and radio among them and the musicians still at work. At the same time, Weber pressed upon the Roosevelt administration his and his union's concerns about the problems of unemployed musicians.[61]

In November 1933 Weber instructed union locals to substitute out-of-work musicians for those then regularly employed in radio and theater orchestras. Specifically, he proposed that bandleaders be retained but that sidemen be rotated "at least every four weeks." He also instructed orchestra leaders to make the rotations in consultation with employers. The proposal brought an avalanche of complaints from employed musicians as well as their employers. Broadcasters insisted on keeping key orchestra members and warned that advertisers would not sponsor bands with substitute players.[62] The management of station KFWB in Los Angeles, to illustrate the response, claimed that the rotation of band members would "upset the whole routine of a radio station," force the station to pay for additional rehearsals, and violate existing labor contracts. At the same time, employed musicians were in no mood to share their jobs and incomes. In the face of these objections Weber soon conceded that his proposal was unworkable, and radio stations continued to hire musicians as they saw fit.[63]

WHILE WEBER succumbed to policies of accommodation, several of the nation's leading bandleaders, among them Fred Waring, Paul Whiteman, and Guy Lombardo, formed the National Association of Performing Artists (NAPA) to increase their own record royalties. These bandleaders, who were also members of the AFM, had the phrase "For Home Use Only" printed on the labels of their recordings, and they sued broadcasters who ignored the labels and played the recordings on radio. This was one of the first attempts by musicians to control the use of their recordings. Like the NAB, NAPA was a response to the growing market power of a countervailing group, in this case broadcasters. Musicians created NAPA to protect themselves against exploitation and thus gain greater benefits from technological change.[64]

The AFM endorsed NAPA's efforts to control the use of records on behalf of musicians. In fact, AFM locals whose members collected record royalties suggested in 1933 that the federation pursue a similar strategy. But the union had no legal standing to sue broadcasters over their use of recordings, since it did not itself render musical services to them. After giving NAPA its blessing, the union watched the new organization and its efforts closely, for the federation had a vital interest in the question of whether broadcasters had to respect contracts between record companies and musicians. If it did, the AFM might be able to protect the rights of musicians in the recordings they made, and thereby gain a measure of control over radio's use of technology.[65]

The test began in 1935, when WDAS in Pennsylvania ignored the inscription "Not Licensed for Radio Broadcasts" on a recording by Fred Waring, and Waring sued. Two years later the Pennsylvania Supreme Court decided the suit in Waring's favor and issued an injunction against unauthorized airing of the bandleader's recordings. Insisting that the law must "adapt itself to new social and industrial conditions," the court disregarded earlier decisions concerning the transfer of property and ruled that the plaintiff's property rights in his recordings did not end with the sale of his records to the public. The court acknowledged that existing copyright laws did not protect Waring but concluded that "the nature of new scientific inventions make restrictions [on the use of recordings] highly desirable." Unrestricted broadcasting of Waring's music, the court found, might injure his reputation as well as his income, especially if stations played outdated recordings.[66]

The decision in the Waring case gave musicians a legal basis for collecting royalties on the commercial use of their recordings. But the ruling was effective only in Pennsylvania, and similar suits in other jurisdictions produced different results. Bandleader Frank Crumit lost a similar case in Massachusetts in 1936, as did Ray Noble in New York in 1937. Elsewhere broadcasters who lost in lower courts won appeals in higher courts. Finally, in 1940, in a case involving Paul Whiteman, a federal circuit court denied musicians the right to collect royalties on the use of their records by broadcasters, ruling that once records were bought and sold, purchasers could use them as they pleased. This decision ended NAPA's efforts in this area.[67]

Despite the failure, the story of NAPA speaks to two basic premises of this book: that the growing capacity to reproduce and disseminate musical performances benefited small groups of instrumentalists to the detriment of others, and that musicians responded to technological change in diverse ways. NAPA represented a handful of popular, wealthy bandleaders whose interests were largely removed from those of rank-and-file musicians. These fortunate few made an unsuccessful effort to increase the already considerable benefits they enjoyed from the sound revolution. Meanwhile, growing numbers of instrumentalists found themselves marginalized and, by the thousands, unemployed as a result of the same revolution. Both groups came therefore to demand that the union take aggressive action against recorded music, even if it meant direct confrontation with the recording industry.

Four

Playing in Hollywood between the Wars

WHILE THE SOUND REVOLUTION eliminated musical jobs across the nation, it did create opportunities in a few media centers, where highly mechanized business firms produced the products that displaced live local talent. As the Great Depression reduced the spending power of the public, scores of ambitious instrumentalists moved to these centers to advance, or often to save, their careers. A fortunate few were able to benefit from capitalist development by relocating to Los Angeles, a growing media center, during the late 1920s and 1930s. Technical innovations in entertainment industries created a completely new work environment there, in which musicians assumed new roles and faced new challenges.

THE BRIGHT LIGHTS of Los Angeles reflected the changing world of working musicians. In the early 1930s theater owners across the city installed new sound systems, thereby displacing pit musicians. As late as 1933 the Paramount, Pantages, Chinese, and Mayan theaters in Hollywood still had sixteen- to eighteen-piece orchestras, while smaller houses such as the Orpheum, Manchester, and Million Dollar employed four- to eight-piece bands. But the popular Loew's State Theater in downtown Los Angeles as well as the houses of Warner Bros., United Artists, and several smaller chains had gone "straight sound."[1]

As old avenues closed, new ones opened. In the early 1930s Los Angeles was a principal production center for the film, radio, and record industries. The city's eight major motion-picture companies produced 85 percent of

American films, and the nation's major radio networks and recording companies relied heavily on flagship stations in Los Angeles. Anomalies in an era of severe depression, these expanding entertainment enterprises created many new jobs. By 1935 perhaps one thousand musicians were working in media industry studios in Los Angeles, the number varying at any one time according to production schedules and other factors. Almost all of these jobs were in the glittering suburb of Hollywood at the foothills of the Santa Monica Mountains, a few miles northwest of downtown.[2]

There, Local 47 of the American Federation of Musicians (AFM) struggled to save theater jobs while trying to exploit new opportunities in film and radio. Organized in 1894, the Los Angeles local had long enjoyed a position of strength in the city's labor movement. In a citadel of antiunionism, and long before the advent of network radio and the talkies, Local 47 had negotiated closed-shop hiring policies in theaters, clubs, and other places that hired musicians. This success was largely attributable to the fact that before the era of recorded music, employers suffered irretrievable losses when musicians went on strike. The union's power in Los Angeles was not unlike that of AFM locals in other big cities.[3]

With the coming of sound movies, Local 47 joined carpenters, painters, electrical workers, and stagehands to bring uniform wages and all-union hiring practices to production sectors of the film industry. The 1926 Studio Basic Agreement recognized five unions of skilled workers and set up a joint labor-management committee to air grievances and settle disputes. Local 47 also worked closely with other unions to secure satisfactory working conditions in other fields of employment. The Los Angeles Central Labor Council, which coordinated union activity in the city, recognized the key role of Local 47 in these and other activities. "All of the labor movement of this city," the council's secretary-treasurer wrote the union's board of directors in 1936, "is conscious of the very splendid co-operation that Musicians No. 47 has rendered to the rest of the movement on every occasion when it has been called upon, and all of the movement has been anxious to find the opportunity to return in some measure at least, the co-operation that you have rendered."[4]

The growing entertainment business in Los Angeles, coupled with the nationwide decline of theater work, made Local 47 the fastest-growing affiliate of the AFM in the interwar years. The local nearly quadrupled in size during the 1920s, to about four thousand members, and over the next ten years, while AFM membership in most large cities dropped notably, the ranks of Local 47 swelled to more than six thousand. The increase should be understood in the context of general population trends in Los

Angeles, one of the nation's fastest-growing cities throughout these years.[5] But the number of AFM musicians in Los Angeles grew at an even faster rate than the local population. Even New York, which supported more band members than any other city in the nation, had fewer union musicians than Los Angeles on a per-capita basis. By 1940 Local 47 was the largest trade union in Southern California and the third largest branch of the AFM, behind only the branches in New York and Chicago, other major media centers.[6]

The influx of musicians from across the nation created problems for the Los Angeles local. Officials realized that the union's future depended on keeping the supply and demand of musicians in equilibrium; the AFM, however, had always recognized the right as well as the need of musicians to travel freely between union jurisdictions. Transfer members therefore expected easy access to local jobs, while resident musicians demanded protection against outsiders. In 1929 Local 47 appealed to the national union for help in dealing with the problem.

In response to the appeal, Joseph N. Weber, president of the national union, addressed the problem at the 1929 annual convention. "Members of the Federation have gone to Los Angeles by the hundreds and have been disillusioned," Weber told the convention, and were now "subject to misery and want." More important, Weber placed the Los Angeles motion-picture studios, and eventually the radio networks too, under the jurisdiction of the national executive board. He also empowered officials of Local 47 to bar transfer members from movie studios for a year, and he put J. W. Gillette, a former president of Local 47, in charge of enforcing the restriction. Gillette filled a new position, international studio representative, the responsibilities of which were independent of Local 47. He soon became known as the "czar of the studios."[7]

The yearlong ban on employment of newcomers in studios discouraged some instrumentalists from moving to Los Angeles. Yet hundreds of depression-worn musicians made the sacrifice for a chance to secure studio work at a later date. They moved to Los Angeles even though union rules prohibited newcomers from accepting full-time work as musicians for three months after their arrival. Many newcomers found part-time work in clubs, hotels, or private engagements. Many also worked outside the music business.[8]

AFTER A YEAR in Los Angeles, instrumentalists could seek work from studio contractors. The contractors, often men with limited musical skills, had agreements with studios to supply orchestras for film production. Through the kind of favoritism this system encouraged, a handful of contractors dominated the market, and the musicians they favored had regular employment. To facilitate the hiring process, contractors kept lists of telephone numbers of available sidemen. For each position in an orchestra, they arranged the names of instrumentalists according to first-, second-, and third-call rank. By the early 1930s, when each of the major motion-picture companies maintained thirty- to forty-piece orchestras, contractors employed about three hundred musicians who worked twenty-five to forty hours a week. They used another one hundred to two hundred instrumentalists on a part-time basis, chiefly when studios augmented their orchestras for major productions or when regular members were absent.[9]

The contractors' control over hiring was a major source of dissatisfaction for instrumentalists, whose employment and income depended on a small clique of insiders. Even when composers or conductors requested individual musicians, as they sometimes did, contractors might ignore the requests. Musicians therefore carefully nurtured relationships with contractors and kept their complaints about the hiring process to themselves. As one instrumentalist put it, "You stand a chance of losing a quarter or half the income for a year if a big contractor, like X, becomes cool to you." Another explained, "You're on a contractor's list and you can be removed from it in a minute." Attitudes toward the contract system might also depend on one's own skill and reputation. Al Hendrickson, a Texas-born guitarist who worked on five thousand films during a remarkable forty-year career, had few complaints about the hiring process. Hendrickson suggested that contractors hired the most capable and dependable musicians. "The guys who did the work over the years," he recalled, "did the job right."[10]

Instrumentalists who benefited from this structure enjoyed some of the best wages and working conditions in the profession. Seated behind music stands with their backs to movie screens and surrounded by hanging microphones and busy soundmen, motion-picture musicians were the envy of other instrumentalists. (In the late 1930s, when public-school teachers earned less than $3,000 a year, sidemen in movie orchestras might make $10,000.) Illustrative of this pattern is the career of Art Smith, a clarinetist and saxophonist who moved to Los Angeles from Caldwell, Idaho, in the 1930s. In 1938 Smith got a job at Disney Studios, but for several months he worked only two days a week, earning a minimum of $30 for each of two three-hour sessions. In 1939, however, he secured steady work at Paramount Pictures, earning $200 a week for five consecutive days of work.

Smith clearly preferred this to other lines of musical employment. "Motion picture work," he later said, "was a marvelous way to make a living."[11]

The experience of violinist Eudice Shapiro provides a different perspective on the character of film work, since among other things it speaks to the status of women in film orchestras. A graduate of the Curtis Institute of Music in Philadelphia and a former student of renowned violinist Efrem Zimbalist, Shapiro moved to Los Angeles as a young woman whose solid reputation had already made connections for her in the film industry. After enduring Local 47's one-year clearance period, she freelanced at Paramount, Universal, United Artists, RKO, and other studios. World War II created new opportunities for her as for many women in other occupations. In 1943 she replaced the outgoing concertmaster at RKO, a job that utilized her ability to solo and to help conductors and composers commu-

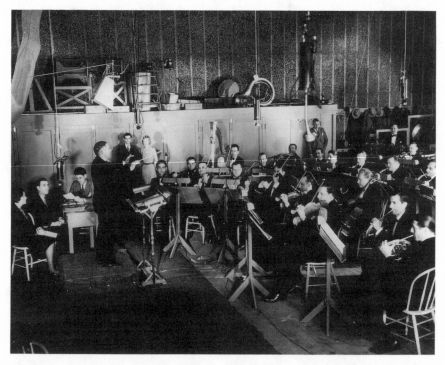

Leo Forbstein conducting the Vitaphone Recording Orchestra at Warner Bros., 1930. Early in his career Forbstein had been a violinist with a St. Louis theater orchestra; at Warner Bros. he headed the music department and pioneered in scoring sound films. *(Hearst Collection, Cinema-Television Library, University of Southern California)*

Stage to Studio

nicate their ideas to instrumentalists. Shapiro was the only female instrumentalist in such a prestigious position in the industry. But in that position she received, in accordance with union rules, twice the wages of sidemen and earned approximately $10,000 a year. Aside from her higher pay, however, Shapiro was treated "just the same as everyone else."[12]

The work was stressful. Producers paying for every wasted minute insisted that instrumentalists perform precisely and efficiently. That fact put a premium on sight-reading skills, for musicians received even the most complicated scores only when they were scheduled to play them. Al Hendrickson admitted that even the best sight-readers worried on some jobs. Hendrickson remembered arriving at one early-morning recording session "just in time for the downbeat" and finding a complicated opening passage written especially for him. "The first cue was a solo that started on the highest fret on the classical guitar." After an uncomfortable delay, he recalled later, "I worked it out, some way. Things like that happened to all of us."[13]

Hard-to-read charts and difficult musical passages put special strains on newcomers struggling to establish reputations. "There's a lot of pressure for the guy just breaking in," one studio musician reported; "the clock is going and you're sitting with a 50-piece orchestra." And if "you can't do it," he added, "there are 50 other guys waiting to have a shot at it." Indeed, making it in the studios required far more than reading charts under pressure. Musicians had to build and maintain intricate webs of informal contacts not only with contractors but with composers, conductors, and an inner circle of leading musicians who could influence contractors' decisions. As one instrumentalist said, "Getting into this jungle is very, very difficult. You have to be very smart."[14]

These and other problems affected the lives of studio musicians in the depression decade. Violinist Louis Kaufman, who began working in Los Angeles studios in 1934, has offered insights into how technology altered the playing techniques of instrumentalists. Studio microphones, he explained, pick up noises that audiences in concert halls do not hear. As a result, Kaufman said, "You have to be a little bit more careful with the bow pressure, you do not dare press and get the extremes of forte that you could get in a hall in which the airspace swallows up a lot of the surface noise." Kaufman noted another difference as well. "The vibrato," he stated, "has to be somewhat heightened, it has to be somewhat faster than you really need for a public hall." He also suggested that playing to microphones made it difficult for musicians to transmit emotion: it was "something of a trick getting around the surface and yet getting the intensity at the same time."[15]

In addition to instrumentalists like Kaufman who worked in large, indoor studio orchestras, a coterie of "sideline" musicians worked in jobs that required them to go wherever scenes were filmed. Their task was to play atmosphere or mood music that inspired actors to accomplish the emotive scenes necessary for successful melodrama. Silent-screen star Blanche Sweet, whose career was apparently ended by the talkies, said that sideline musicians had a major role in the production of her movies. Their music, she recalled, "seemed to help everyone from the stars to the technicians, stage hands, carpenters, electricians, everybody." Colleen Moore, who starred in the film *Irene* in 1926, agreed: "We always had mood music on the sets. I had a three-piece orchestra that played continually, not only to put us in the mood but to amuse us between scenes, since I was making comedies and needed to keep in high spirits." But not all moviemakers used sideline musicians. Moore recalled that the renowned director D. W. Griffith, who made films until 1931, "never used any music while he was filming. He always said that he would never employ actors who could not feel the role enough to weep at rehearsals."[16]

Sideline work occasionally led to acting jobs, since union rules encouraged moviemakers to use union musicians in music-playing roles. In such cases producers sent musicians to local costume companies to be dressed appropriately. Civil War uniforms, Roman armor, and western clothes were among the outfits musician-actors wore. Despite the possibility of bit-part appearances in motion pictures, many musicians disdained sideline work, since the music sideliners played seldom appeared in film soundtracks. Yet union pay scales for sideline musicians were comparable to those of studio musicians. In the late 1930s sideline instrumentalists made $15 to $40 a day for work that was intermittent as well as undemanding. "On sidelines," one musician recalled, "I spent most of my time playing cards and reading books."[17]

Closer to the top of the music hierarchy in the film industry were composers, whose positions were not unlike those of staff employees of the studios themselves. Composers usually worked frantically for several weeks after the end of filming, matching music and film scenes. In doing so in the early 1930s, they used small machines called Movieolas that showed the film and a "click-track" that helped them coordinate music and movie sequences. However glamorous it seemed to outsiders, the task of putting film to music was tedious and nerve-racking. In 1945 Ernest Gold recalled

Sideline musicians on location, 1927. Movie producers often hired instrumentalists to inspire actors and actresses during the filming of emotional scenes. Here musicians await cue during filming of *Johnny Get Your Hair Cut,* a Metro-Goldwyn-Mayer production starring Jackie Coogan. Seated left to right are Coogan, Coogan's father, and MGM executive Joel Engel. Director B. Reeves Eason stands to the right of the cameraman. *(Bison Archives)*

his first experience as a film composer. "When I arrived at the studio that afternoon I was given a stopwatch, pencils and paper, and the cue sheets," he said. "I was also told I was only allowed nineteen men in the orchestra since it was a picture with a small budget." After completing the score— twenty-five minutes of music—in five days and nights, Gold learned that new footage had been added to the film and that he must revise the music within a limited time. "I do not intend to work at this breakneck speed again," he said of the experience, "nor do I recommend anybody to do so."18

Composers typically had musical assistants called orchestrators, who were seldom bound to a single studio but worked instead for composers, writing and rewriting scores for individual instruments. Copyists made legible scores of their writings for orchestra members and in doing so altered particular passages as problems arose in the final stages of production. A separate musical director might coordinate all of this activity, though some composers acted as their own musical directors and even as orchestra leaders. In the 1940s such major composers as Max Steiner, Adolph Deutsch, Alfred Newman, and André Previn filled several of these roles and earned several thousand dollars per film.[19]

The early career of Austrian-born Max Steiner, head of RKO's music department in the late 1920s and 1930s, personalizes the experience of film composers in these years. In 1929 this former New York vaudeville pianist turned theater bandleader was conducting an orchestra in Boston when the production chief of RKO offered to make him musical director of the entire studio at a weekly salary of $450. Steiner did an outstanding job. His best-known work at RKO, the now-classic film *King Kong,* starring Robert Armstrong and Fay Wray, appeared in 1933. Working night and day for eight weeks, Steiner achieved a sense of realism in this highly unrealistic movie about a prehistoric beast struggling to survive in a modern urban setting. With simple themes distinguishing the leading characters and musical passages that heightened the drama at critical moments, Steiner's music added tension and intensity to the experience of seeing the film. It underscored the emotions of terror, loneliness, anxiety, and love that the film's director sought to evoke. The eerie melodies of harps as the boat approached Skull Island, the descending three-note motif that identified the monster, and the frenzied crescendos of strings, cymbals, and drums that accompanied his fall from the Empire State Building showed the power of music in film. Even the producer, Merian C. Cooper, agreed that much of the movie's popularity was due to Steiner's music.[20]

Ironically, movies made during these years, even those with musical themes or subjects, concealed the crisis facing musicians. Many films made in Hollywood's Golden Age gave the impression that musicians were a fully employed, happy-go-lucky group. The many musicals of the early 1930s gave an especially deceptive image of instrumentalists. *King of Jazz,* a 1930 Universal film about Paul Whiteman and his forty-piece orchestra, for example, showed musicians working in lavish settings and wearing the finest clothes. Audiences heard singer Bing Crosby speak of the "higher, fine things of life" while they watched band members in white tuxedos and top hats playing grand pianos in the most opulent of settings. Members of

Whiteman's orchestra may have enjoyed what Crosby called the "silver lining," but the average musician was more likely to be down and out.[21]

Throughout the 1930s all three radio networks—NBC, CBS, and Mutual—broadcast programs from Los Angeles. Their stations there provided lucrative full- and part-time work for a few hundred talented, and fortunate, instrumentalists. A half-dozen smaller stations in the city, each of which occasionally carried network programs, also employed orchestras for live broadcasts. Altogether, in 1935 the industry employed perhaps four hundred musicians in Los Angeles. The executive board of the AFM handled labor negotiations with the networks, but Local 47 regulated the conditions of work.[22]

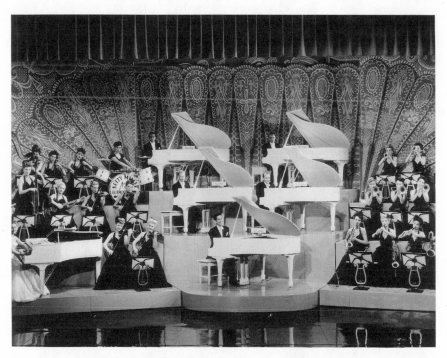

Scene from *Out of This World*, Paramount Pictures, 1944. Hollywood filmmakers typically portrayed musicians as happy-go-lucky and fairly well paid at a time when growing numbers of them were down and out. Here an "orchestra" of "glamour girls"— only two of whom were professional musicians—played accompaniment to the famous pianists Carmen Cavallaro, Ted Fiorito, Henry King, Ray Noble, and Joe Reichman. *(Courtesy of Jerry Anker)*

The local classified network orchestras as "sustaining" or "commercial." Sustaining orchestras, typically of eighteen to twenty-five pieces, contracted to work five or six days a week for forty or fifty weeks a year on "nonsponsored" programs. Their members were known in the industry as staff musicians. Sustaining orchestras apparently became more and more versatile during the 1930s. They played classical as well as popular music, often backed up by well-known singers, and even played "bridges" and "cues" for dramatic programs or comedy shows. One study of music in early radio notes that staff musicians "might be called upon to accompany a classical singer, to glide through a lilting Strauss waltz or to perform a rousing Sousa march."[23]

Violinist Lenny Atkins later recalled his work in a sustaining orchestra in Los Angeles during this period. "Every day was different," he remembered; "sometimes we worked twelve hours a day, sometimes we worked three." He might be called to work anytime between eight A.M. and four P.M., and he spent most of his time rehearsing for live broadcasts. Although instrumentalists often tired of playing familiar songs, they seldom complained about their work.[24] The federation required radio stations to hire staff orchestras of a minimum size on a yearly basis. This meant that musicians in sustaining orchestras enjoyed the security of guaranteed incomes and two weeks' vacation as well. Their weekly earnings of $100 to $120 meant annual incomes of $5,500 to $6,500, compared with an average of $1,500 for skilled factory workers at the time.[25] Commercial orchestras worked differently. Usually fifteen- to twenty-piece groups hired to serve weekly sponsored programs, they played for several shows each week. They also played "intros" and "themes" for talk-oriented programs and provided music to back up well-known singers and musicians. Some commercial programs featured celebrity traveling bands stopping in Los Angeles for weekly shows. The Kay Kyser Orchestra played in cities coast to coast, for example, but returned to Los Angeles once a week for NBC's *Lucky Strike Program*.[26] The experience of Henry Gruen, a saxophonist who moved to Los Angeles in 1938 from San Antonio, shows one side of the life of musicians who worked part-time in network commercial orchestras. In 1941 Gruen secured a job in Ozzie Nelson's fifteen-piece band at NBC. He arrived for work at five P.M., then spent an hour and a half rehearsing for a live broadcast that lasted from seven to seven-thirty. At a time when a pound of good steak cost 15 cents and a loaf of bread 10 cents, Gruen earned $25 to $35 per program.[27]

Local 47 required small network-affiliated stations to hire musicians on a full-time basis. The union yardstick for orchestra size was still station wattage: the more powerful the station, the larger the orchestra must be. Thus, KFWB had to have an eight-piece band while KMPC got by with only four musicians. Wages at these stations were 10 to 30 percent below those at network stations, and the musicians they employed performed in sponsored as well as nonsponsored programs.[28]

Hierarchy among musical workers in radio varied according to the size and value of the stations they worked for. In all stations the orchestra leader occupied a position of authority over other musicians. Leaders often contracted for instrumentalists as well as conducted the orchestra, and they earned at least twice as much as the highest-paid sidemen. Their responsibilities justified the differential. Leaders decided the arrangement and orchestration of songs as well as the tempo and moods, and they performed various administrative duties. At some stations they dealt with sponsors and program directors in planning musical shows, and everywhere they provided sidemen with musical scores, which meant they worked with copyists and arrangers as well as union stewards. Larger stations employed orchestra managers or musical directors to assist the leaders, though leaders sometimes hired their own assistants. Popular bandleaders like Fred Waring, Guy Lombardo, and Paul Whiteman worked in several capacities and received especially high salaries.[29]

Program directors had final authority in matters of production. These were men with the technical skills and knowledge necessary to oversee the production of various kinds of programs. Some had backgrounds in music; others, in acting, writing, or radio announcing. The general responsibility of program directors was to ensure that shows had unity, balance, and quality. The directors worked closely with studio performers, especially orchestra leaders. During rehearsals they timed the parts of programs, arranged the position and volume of microphones, and otherwise worked to make certain that the live performances succeeded. When programs aired, it was their job to see that they began and ended on time, "on the nose." During performances directors relied on standardized hand signals to communicate with bandleaders. An outstretched arm with pointed finger, for example, told leaders to begin the program, while a finger rotating clockwise meant to speed up the tempo. Other signals instructed the leaders to move closer to the microphone, fade out the music, or stand by for upcoming cues.[30]

The program director gave these signals from inside a control booth, the nerve center of the production. These small rooms, found in all Hollywood studios, varied in size and shape but not in function. The booths housed many of the technologies that enabled broadcasters to regulate what went on the air. Sound flowed into and out of the booths via electrical circuitry, though they were thought of, paradoxically, as soundproof because their design blocked out the sounds of the performers. The most prominent feature of every booth was the control, or mixing board, equipped with volume controls for each microphone, a master volume control, and various dials measuring volume levels. Large overhead clocks helped the director and engineers check the timing of the show, and glass windows, loudspeakers, and microphones helped them see, hear, and communicate with studio performers. In large stations live performances flowed from these control booths to a master control room, which distributed the programs to transmitters and network lines.[31]

The language of the studios throws light on more than the production process, for it graphically illustrates what had happened in the musicians' world. Control over music production had shifted from the stages of theaters and the floors of dance halls to small enclosed booths in industry studios. The shift created a much more absolute control by transferring power from workers to management. In the old work setting, direction of music was in the hands of "leaders," men who by virtue of their own musical skills "led" other musicians in the production process. In the automated studios, in contrast, control of production was centralized in aptly named "control" booths, where men who were not themselves musicians signaled commands to musicians according to the measurements of technological instruments. The language they used no doubt affected the self-perceptions of both management and labor. If "master" controllers directed performances from "control" booths, what did that make of musicians? Had they become minions, or even automatons?

FILM AND RADIO orchestras in Los Angeles between the 1920s and early 1940s included very few African Americans or Hispanics. Reedman Art Smith, who worked in both film and radio at that time, said later, "I would have been shocked to have seen a black musician in the studios." One historian has suggested that the studios were hesitant to employ black bandleaders, and thus the sidemen in their bands, for fear of offending white audiences. In fact, Cab Calloway, emcee of a prime-time network radio show in the early 1940s, and pianist-composer Duke Ellington, whose music graced several Paramount films, were two of the few African Americans who worked in radio or film on a regular basis. Among the few Mexican Americans who found work in studios were the members of Los

Madrugadores, a musical group that performed during early-morning hours, from four to six A.M., on stations KMPC and KELW (Burbank). Minorities generally found studio jobs only in productions made specifically for minority audiences.[32]

Like the rest of America before the civil rights movement of the 1960s, Los Angeles was racially segregated during the interwar years. African Americans lived and worked mostly in the southern parts of the city, Mexican Americans in the east. In fact, black musicians in Los Angeles had organized their own union by 1920. A decade later, the all-black Local 767 had about two hundred members as well as its own headquarters and its own staff of business agents to police clubs and restaurants in which black instrumentalists performed. To protect each other, the white and black locals had a common wage scale. According to John TeGroen, vice president of Local 47 in the 1940s and president when the two locals amalgamated in 1952, relations between the two unions were always amicable. Local 767, he said, never demanded greater access to studio work, though the absence of civil rights laws no doubt encouraged this reticence.[33] One place where African Americans might have complained about segregation and its consequences was the national conventions of the AFM, but in fact few black locals, including Local 767, sent delegates to the conventions before the 1930s. Race relations, especially in the Deep South, made things difficult for those that did.[34] The lack of influence at national conventions that these relations created weakened the voice of black musicians within the AFM and thus helped perpetuate some of the problems they faced.

If the near absence of black and Hispanic musicians in film and radio work was largely a matter of racial discrimination, the fact of the absence shed light on basic matters of musical skill. Producers and directors in film and radio placed far less value on improvisation and individuality of interpretation, musical qualities commonly attributed to black and Hispanic instrumentalists, than on sight-reading skills. In other words, the technical changes that gave rise to new opportunities in media centers also encouraged specific definitions of virtuosity. Skills that worked to advantage in club or theater work, or even in record production, did not have the same advantage in film and radio work. Still, the paucity of minorities reflected the importance of social acceptability in studio employment. The oversupply of instrumentalists allowed bandleaders to be highly selective in staffing film and radio orchestras. With so large a pool of available talent, they could and did use personal and social factors, including race, as a basis for hiring or refusing to hire individual musicians.

Minorities were more visible in the city's expanding recording industry.

Los Madrugadores made dozens of recordings during the depression decade; one member of the band estimated that the group recorded over two hundred discs for seven different companies, including industry leaders RCA-Victor, Columbia, and Decca. Black musicians found recording work as well, mostly in fledgling independent companies specializing in blues recordings. The deepening pool of talented entertainers in Los Angeles combined with technological developments in record production to spur the growth of the city's recording industry throughout this era.[35] Locals 47 and 767 had strict guidelines for the employment of instrumentalists at record companies. The companies hired musicians not on a staff, or full-time, basis, but by the recording session. In the late 1930s instrumentalists earned at least $24 for a two-hour session and at least $6 for each additional half-hour. For radio transcriptions, which were often simply recordings of live broadcasts, they earned $18 an hour.[36]

Regardless of race or ethnicity, most musicians who worked in studios supplemented their incomes by working in clubs and hotels. Some studio players, especially part-timers, earned the bulk of their income in such places. Many instrumentalists worked on Sunset Boulevard in clubs like Ciro's, the Mocambo, and the Trocadero. The Venice Ballroom in nearby Santa Monica, the Biltmore Bowl in downtown Los Angeles, and other such places hired house bands of eight to ten pieces to perform six or seven nights a week. The city's largest dance clubs were the Palomar Ballroom on Vermont Avenue and the Palladium on Sunset Boulevard, where the biggest bands played. Dozens of cozier places featured trios or quartets.[37] Musicians typically secured casual work through bandleaders who had agreements with proprietors. Although bandleaders actually paid their sidemen, Local 47 enforced minimum wage scales and maximum hours for casual musicians. Proprietors who violated those standards found themselves labeled unfair and denied the services of union musicians, while musicians who violated the union sanction were fined or suspended from the union.

During these years broadcasters, using microphones and telephone wires, began to pick up musical performances at hotels and dance clubs and transmit them across the country. One example was the popular broadcasts from the Coconut Grove at the luxurious Ambassador Hotel on Wilshire Boulevard, where the bands of Artie Shaw, Ben Bernie, and Anson Weeks played regularly. The management of such locales were eager to permit the broadcasts in return for the free advertising, while the musicians did not complain of the free exposure. Yet instrumentalists saw such broadcasts as a threat to their jobs, and their union sought to control them. Thus, Local 47 required hotels, theaters, and dance clubs to pay their instrumentalists 25 to 35 percent above union scale when they permitted remote broadcasts. It waived the surcharge, however, if the station carrying the remote broadcast employed a staff orchestra.[38] Such union-imposed rules reflected the balance of power between the union and management in the music business. They also reflected the fact that workers as well as managers could use technological innovations to advantage.

Much of the music that played in Los Angeles during the 1930s was the energetic and lighthearted tunes of the swing era. Orchestras filled dance floors with the rhythmic beat of Glenn Miller's "In the Mood" or Benny Goodman's "Stompin' at the Savoy." After a few upbeat songs couples often requested soft ballads such as Hoagy Carmichael's "Stardust" or Jerome Kern's "Smoke Gets in Your Eyes." At the Ambassador Hotel, pianist Gus Arnheim and his band opened each performance with a smooth rendition of Irving Berlin's "Say It with Music." Other bands introduced the melodies of Cole Porter, George Gershwin, and Rodgers and Hart. In other nightspots such as the Cotton Club in Culver City, audiences preferred the less structured improvisations of jazz played over the chord progressions of Duke Ellington, William "Count" Basie, and Earl "Fatha" Hines. Across the city, music pulled much of the public into a lively nightlife. Los Angeles became a city oriented toward entertainment—as business as well as pleasure.[39]

A closer look at the character of casual work underscores the uniqueness of the workplace of most musicians. Casual bands usually worked in dimly lit, crowded places late at night. They performed before live and sometimes rowdy audiences who expected a wide variety of popular songs played with a traditional sense of sequence, unity, and finality. This contrasted notably with the situation of studio musicians, who typically performed in bright, spacious settings during daylight hours. In addition, the latter often played for people hundreds of miles away and even further removed in time. Also unlike casual bands, studio orchestras performed disjointed strains of music designed to match the changing scenes in films or the unfolding plots of radio programs. Differences in work environments, in other words, went far beyond matters of wages, hiring patterns, or divisions of labor.

The contrast in work settings had important implications for the culture of the workplace. Many places where casual musicians worked, even before the end of Prohibition in 1933, encouraged the consumption of alcohol. One result of this was that musicians often drank on the job. The permissive underground atmosphere of some clubs and dance halls also encouraged the use of narcotics. Some casual musicians drank or used drugs to be sociable with customers or proprietors; others did so to relieve stress and enhance their performance. After smoking marijuana, one musician explained, "I felt I could go on playing for years without running out of ideas or energy. . . . I began to feel very happy and sure of myself." For a variety of reasons, however, the use of alcohol and drugs was unusual in studios. Daytime hours of operation and the absence of bars partly explain this. But trumpeter Bob Fleming, who worked at MGM in the late 1930s, suggested that alcohol and drugs were incompatible with the nature of studio work. "Musicians who were drinking or on drugs," Fleming explained, "were not apt to be great sightreaders. We couldn't just go into a jam session, or go into jazz." The fact that studio musicians had substantial income at stake if they fell out of favor with contractors also encouraged sobriety in the workplace.[40]

The technologies that gave rise to high-paying jobs made possible a distinctive lifestyle for the most affluent musicians. While instrumentalists in casual bands typically rented homes in low-income neighborhoods and drove inexpensive automobiles, growing numbers of fully employed studio players purchased large houses in posh neighborhoods and new cars to match their surroundings. Many also developed a fondness for golf, and their numbers were sufficiently large to lead the trade paper of Local 47 to cover city golf tournaments and keep readers informed of how well, or poorly, union members played. How musicians lived, as one instrumentalist explained, "depended on each one's attitude, and [on] their good or bad habits."[41]

MUSICAL EMPLOYMENT in the Los Angeles radio industry increased notably after 1936, when AT&T altered a longstanding pricing policy. Previously the phone company had billed broadcasters for use of telephone lines that connected network affiliates in a way that encouraged the production of network programs in or near New York. Under that system the company calculated fees on the assumption that radio signals went first to New York, whence they were relayed elsewhere, as had indeed been the reality when radio began. By charging networks by the mile from the point at which a radio program originated to its transmission facilities in New York City and then adding a mileage fee from New York to the point of reception, AT&T made it costly to produce programs on the West Coast.

Labeling this "double rate" pricing procedure an "unnatural" trade barrier, President Franklin D. Roosevelt authorized a federal investigation of AT&T in 1935. The company promptly eliminated the procedure, and the number of radio shows produced in Los Angeles increased. This political intervention against a monopolistic business practice had important implications for Los Angeles musicians.[42]

The new opportunities it generated brought with them new problems. The incomes of those in position to seize the opportunities rose impressively, but more and more musicians were frustrated by their inability to get studio work. The growing disparity between rich and poor musicians divided instrumentalists into quarreling factions. Underemployed instrumentalists, the larger of the factions, complained that a few contractors monopolized the work and fixed orchestra performances so that a handful of sidemen and bandleaders worked around the clock. A single contractor, they pointed out, managed eighteen commercial radio shows. They also complained of the practice of paying overtime wages to fully employed musicians while so many others were unemployed or underemployed. Contractors justified overtime wages by explaining that production problems arose suddenly and required quick fixing; they could not, they insisted, predict when, or how long, their employees might need to work overtime. Yet there was more to the problem than that. Musicians as artists were not interchangeable entities. Regularly employed musicians were known quantities, highly skilled and dependable. Replacing them on short notice with musicians of unknown quality and reliability was a risk most businessmen were unwilling to take when they did not have to.[43]

Another source of grievance for unemployed and underemployed musicians was broadcasters' growing practice of inviting well-known bandleaders and their bands from outside Los Angeles to come perform on network commercial programs. To help make their invitations more alluring, broadcasters worked with talent agencies to book the outside bands in clubs, hotels, and cafés during their stay in Los Angeles. When Jimmy Dorsey and his orchestra were in Los Angeles to perform on NBC's *Kraft Cheese Program* in the mid-1930s, for example, they also played at the Palomar Ballroom. That engagement deprived thirty-three local musicians of at least one night of employment.[44]

As outside groups grabbed more and more business, local musicians proposed that employers who hired outsiders pay the union a "standby" fee equal to the wages of the musicians displaced. In June 1936, after bands from New York replaced local orchestras on the weekly shows of Jack

Benny, Bing Crosby, and George Burns and Gracie Allen, officials of Local 47 asked the national union for help in resolving this problem. Cliff Webster of the local told the president of the national union that the anger of local musicians at the influx of outsiders had grown "to the point of actual threats of personal violence. . . . When men see their bread and butter taken away from them by outsiders, they do not reason things out."[45]

To quiet the rising tide of protest, Local 47 and the international executive board acted to stem the influx of outsiders. In July 1936 the executive board ruled that conductors entering a jurisdiction to work at one establishment could work at no other establishment in the jurisdiction. A month later the federation ruled that studio orchestras must be composed of local musicians, and that orchestras under contract to play regularly for the networks could engage in other work only if they paid the union half of their earnings from the other jobs.[46]

In 1937 the international studio representative of the AFM in Los Angeles adopted additional measures to prevent a handful of instrumentalists from monopolizing studio work. Musicians employed in sustaining bands could no longer play on commercial radio programs, and those who made more than $77.50 a week in one studio could not work in other studios. Like the federation's earlier efforts to spread the work in theaters and radio stations, these rules caused constant headaches for local officials. "It's hard to tell a man he can't make a buck," said John TeGroen, vice president of Local 47.[47] Clearly the new rules penalized full-time musicians, and musicians in heavy demand protested it. But the union insisted on the new rules, which shaped patterns of studio work until the courts outlawed them in the early 1950s.

THE RISE OF studio musicians and the elimination of theater orchestras were two parts of the same process. Mechanization in the music business drastically reduced total musical employment but created unprecedented opportunities for some instrumentalists in Los Angeles and a few other media centers. The fact that business firms depended heavily on the skills and reliability of the musicians they employed gave Local 47 considerable clout in industrial relations, but the influx of outsiders saturated the market and undermined job security. The resulting competition splintered instrumentalists as workers, and their union could only meliorate, not eliminate, the resulting discord.

Five

Rising Militancy

FOR NEARLY A DECADE after the introduction of sound movies and network radio, American Federation of Musicians (AFM) president Joseph N. Weber had urged musicians to adapt to changing industrial conditions by arguing that resistance to technological change would fail. At the same time, Weber and other AFM leaders had tried to control the rapidly changing work environment of musicians. But in the late 1930s the union changed course and pursued bold new strategies. Unlike the contemporary struggles of workers in the steel, rubber, and auto industries, the musicians' struggle attracted little media attention. Nonetheless, it was real and important. Like the other disputes, it reflected growing concern about the maldistribution of wealth in America and showed that musicians, like other workers, could act to protect their jobs and income. While the union was striving to protect the interest and well-being of musicians in the half-decade or so around 1940, businesses too were facing their own challenges from technological change and fighting their own battles for survival.

THE IMPACT OF the sound revolution continued in the latter half of the 1930s, when approximately three thousand more theater musicians lost their jobs. Meanwhile, network programming and recorded music continued to reduce local opportunities in radio. In 1937 Weber complained that barely 10 percent of the nation's six hundred commercial stations employed musicians, together generating fewer than eight hundred full-time jobs.[1]

Other industrial developments presented additional challenges. Jukeboxes continued to displace instrumentalists wherever people wanted to hear music while they ate, danced, or otherwise relaxed. By the summer of 1937 roughly two hundred thousand of these music machines were operating nationwide, and entrepreneurs were installing more of them every day. A brisk trade in secondhand machines had curtailed production, but companies like J. P. Seeburg, Wurlitzer, and Rockola manufactured eighty thousand new machines that year. Many operators who installed and serviced the machines handled more than one hundred of them in retail locations and maintained stable, mutually beneficial relationships with proprietors who provided them to music consumers. The ingenious machines afforded millions of people hours of pleasure at nominal cost while indirectly costing musicians hundreds, even thousands, of jobs.[2]

"Wired-music" services were a newer mechanical menace. As early as 1907 the Telharmonic Company in New York had transmitted recorded music over telephone lines. The company failed to make commercial use of the transmission it pioneered. However, a similar firm, the Wired Music Company, appeared in 1931 and made headlines by picking up the music of a three-piece nonunion band at New York's Barclay Hotel and piping it to more than fifty local hotels and restaurants. The future of musicians in hotels and restaurants became even more problematic in 1934, when the North American Company established a new subsidiary, the Muzak Corporation. The federal government soon forced North American to divest itself of the subsidiary, which came under the control of Associated Music Publishers, a group sympathetic to musicians. Associated Music agreed to deploy Muzak technology without jeopardizing musical employment, but other wired-music companies made no such agreement.[3]

Even bandleaders who benefited from new recording and broadcasting technologies had reason to oppose these developments. The elite among musicians, these bandleaders found themselves victims of bootleggers and pirates surreptitiously stealing their live performances from radio broadcasts. Amendments to the Communications Act of 1934 had outlawed that practice, but musicians continued to complain about hearing their own radio music on records they had not recorded. At the same time, movie musicians protested that independent film producers were taking old soundtracks from music libraries and paying a few musicians to rerecord or dub new musical passages onto them, thereby eliminating the cost of orchestras.[4]

Meanwhile, AFM membership beyond the nation's largest media centers remained stagnant. In 1936 many locals were still 20 to 35 percent smaller than they had been in 1928. Those in Baltimore, Minneapolis, Atlanta, and some other cities were about half their former size. Hard times continued to take the heaviest toll on the union's smallest locals, many of which folded in these years. The total number of chartered locals in 1936 stood at 641, the lowest number since 1913. The fact that membership was rising in New York and Los Angeles was of little comfort to musicians and union officials living elsewhere. Indeed, it only increased their frustration.[5]

Concern over these and other problems made the convention in Louisville, Kentucky, in 1937 one of the most clangorous in AFM history. Delegates demanded that Weber launch an aggressive campaign against the "undue use" of recorded music and insisted that unless he stimulated musical employment, he should call emergency sessions of all locals to work out ways to fight canned music. Some locals wanted an outright ban on recording. "It is absurd," delegates from New Orleans Local 174 maintained, "for musicians to furnish the music for making sound films." The New Orleans delegates then asked Weber to prohibit musicians from recording until record manufacturers agreed not to sell their products to broadcasters and jukebox operators. New York Local 802, meanwhile, presented fifteen specific proposals to restrict the mechanization of music. "The abuse and misuse of mechanical reproductions of music," the New York delegation warned, "constitute a threat which may annihilate the profession."[6]

These calls for action represented more than dissent from the policies of Weber and the AFM; they reflected growing disaffection among the rank-and-file musicians and demands for more militant strategies. Faced with shrinking job markets in a persistently depressed economy, musicians across the country were demanding that the union face up to the crisis. In 1937 a disgruntled member of New York Local 802, J. F. McMahon, wrote to William Green, president of the American Federation of Labor (AFL), berating Weber's policy of accommodation. "[Weber] refuses to assist us in our drive for Live Music," McMahon complained; "[he] says it is futile. . . . We must face facts. Invention, mass production, and unfettered monster trusts have us by the throat, [but] Weber does nothing." McMahon thought Weber "useless" and even a "hindrance" and suggested that Green investigate Weber's use of union funds while encouraging musicians to remove him from office.[7]

Union leaders realized that such surging unrest jeopardized their positions. That is perhaps why the 1937 convention redefined the way the union dealt with technological innovation. For nearly four decades the federation had opposed every suggestion to boycott record manufacturers or those who used their products. Now union leaders appeared ready to change that stance if they could find a way to deal with so intractable a problem.

Clearly sound films, records, and network radio were here to stay. Accepting that premise, Weber and the union executive board decided that radio, whose growing profits during the depression defied national economic trends, held the key to the musicians' dilemma. In closed-door sessions the board agreed on a plan to pressure the networks to increase the number and size of the staff orchestras employed by their affiliated stations. The strategy reflected the fact that there were many more affiliated than network-owned stations, and those stations therefore had the potential for creating far more jobs than did the networks themselves, and jobs too across the nation. The union thus threatened to strike the networks and record companies—the sources of most of the material the affiliates broadcast—and hoped the threats would cause the networks and record companies to force the affiliates to accept union demands to increase employment in radio. The same action might also increase employment in nonaffiliated stations, thus easing somewhat the problems caused by jukeboxes, wired music, and pirating.

On July 13, less than a month after the Louisville convention, Weber set this plan in motion. In sharply worded letters to network executives he threatened a strike unless they agreed to union demands. "Kindly be advised," he wrote, that members of the AFM "will cease to render services at any broadcasting station" on August 14, 1937, "unless [you] have agreed to the regulation of the indiscriminate use of phonograph records or electrical transcriptions." Weber invited the executives to meet him in New York to discuss ways to prevent the strike. "If you fail to respond to this invitation," he stated, "then you will, of course, leave the Federation no other alternative except to hold that your corporation, organization, broadcasting system, individual station or network is no longer interested in having services rendered to it by [AFM] members."[8]

Three days later, in an interview published in *Variety*, Weber elaborated the union's concerns. Denouncing the programming policies of affiliated stations, he insisted that musicians would no longer work for networks that fed programs to stations without staff orchestras. His meaning was clear: unless radio made concessions on this matter, the union would strike the industry. He also hinted that it might strike the record industry too.

The latter threat became explicit when New York Local 802 publicly instructed its members to make no phonograph records or electrical transcriptions after September 30.[9]

In response to these threats, leaders of the recording and radio industries agreed to meet with union leaders. Thus, on July 26–27 Weber and the union executive board met with representatives from leading record companies, among them World Broadcasting, Trans-American, RCA-Victor, Brunswick, and Decca. In two subsequent meetings, on July 29 and 30, they also met with representatives from the major networks, the National Association of Broadcasters (NAB), and several affiliated stations. In a clever move designed to show that he had the support of another union, Weber held some of these meetings in the New York headquarters of the American Society of Composers, Authors, and Publishers (ASCAP).[10]

These dealings between the union and the executives and legal staffs of some of the nation's largest corporations underscored the revolutionary changes that had recently occurred in the music business, and thus in the work environment of musicians. The array of business talent, legal expertise, and financial power Weber and the union confronted showed that instrumentalists and their union now had to deal with sophisticated, well-financed corporate enterprises committed to minimizing production costs through new technologies and new strategies in labor relations, and fully capable of achieving their purpose. The task of the men facing Weber and his union across the bargaining table was to prevent the union from controlling—or even meaningfully influencing—the circumstances, economic or otherwise, created by the new technologies. The AFM might wrest specific concessions from such men, but its chance of dominating the negotiating process was small indeed.

It is worth noting that the corporate spokesmen in this confrontation represented a distinct social group. They shared common ethnic and class backgrounds, and thus educational and career experiences. They even had similar lifestyles. Indeed, homogeneity in the business community went far beyond matters of race, ethnicity, gender, and class. It also involved considerations of ideology and world-view. Corporate policymakers shared common assumptions about the appropriate way to organize industry and society. Their assumptions embraced notions of the sanctity of private property, individual initiative and responsibility, and the subordination of labor to management. While reproving state interference in their own affairs, they nevertheless expected government help in overcoming the problems they encountered in the production and exchange of goods.

Most of them had come to accept trade unions as a fact of industrial society, yet they continued to view concrete union demands as assaults on managerial prerogatives. This, in brief, was the corporate psyche AFM officials confronted toward the end of the depression decade.[11]

IN THE NEGOTIATIONS that resulted from the union demands, Weber told industry leaders that musicians intended to strike the radio and recording industries because they had no choice. The high unemployment among instrumentalists—thirteen thousand AFM members, he noted, were on public relief—was forcing the union's hand, and one of the "prime reasons" for this situation was "the uncontrolled use of recordings." "[Records] are multiplied and duplicated by the thousands," Weber complained, "and then sold to and used in places where otherwise musicians would be employed." In an appeal to the better side of recorders, he spoke of the cultural implications of these practices. He argued that the practices had so reduced the incentives for young people to enter the music profession that the music business itself would soon have to rely on artists of inferior talent. "If musicians cannot find employment," Weber warned, "good musicianship will reach a lower level and the rendition of good music will become the exception rather than the rule." The decline of the nation's musical culture, he predicted, "is only a matter of time."[12]

Weber offered various proposals for regulating the commercial use of recordings. The most important and controversial of these would prohibit recorders from selling recordings to businesses that profited from playing music but did not employ musicians, which meant jukebox operators as well as network affiliates without staff orchestras. To enforce the prohibition, the union proposed to license recordings and through the licensing process restrict the use of recordings. Manufacturers who failed to license their recordings, or who violated the stipulations in the license, would be labeled unfair and denied the services of union members. Industry executives quickly branded Weber's proposal a secondary boycott that violated the antitrust laws and constituted a restraint of trade. Any effort to implement the proposal would therefore spawn expensive legal battles.[13]

Again the union had run up against the antitrust laws. The original antitrust law, named for its sponsor, Senator John Sherman of Ohio, ostensibly ensured competition in the marketplace by outlawing monopolistic practices of big business. The law, however, did not define *monopoly* or specify the monopolistic practices it forbade. It was therefore a vaguely worded endorsement of the antimonopoly principle and dependent for its meaning on judicial interpretation. The implications for labor became clear in 1894, when a federal judge in Chicago found striking railroad workers to be a combination in restraint of trade and enjoined them from interfering with the operation of the railroads they were striking against. Over the ensuing decades other courts applied that precedent to many striking workers. Even the Clayton Antitrust Act of 1914, which specifically exempted labor unions from the provisions of the Sherman Act, did not prevent courts from using the antitrust laws against union strike activity.

Despite the threat of the antitrust laws, AFM leaders defended their proposal to license and restrict the commercial use of recordings. When an industry leader mentioned the Sherman Act, Weber retorted, "A worker has a right to stipulate under what conditions he wants to work, where he wants to work, and how he wants to work." James C. Petrillo, already an influential member of the union executive board, echoed Weber's sentiments: "Does this gentleman know any Federal law, or city law, or state law," Petrillo asked, "whereby a man cannot protect his livelihood?" Other board members, including Charles Bagley, vice president of the union, and Chauncey Weaver, secretary of the executive board, responded to questions with equal assertiveness, especially questions challenging union figures on unemployment among musicians. Finally, at the end of a two-day bargaining session, attorney Milton Diamond of Decca Records suggested that the union give recorders time to consider the union proposals and, if appropriate, formulate counterproposals that the two sides could then discuss. Weber agreed to the suggestion, and the negotiations were adjourned.[14]

When AFM leaders met shortly thereafter with representatives from radio, Weber reiterated the problems of musicians and offered fourteen proposals for solving them. Some of the proposals were similar to those made earlier to recorders. Others asked the networks to expand the size of their staff orchestras, end the practice of recording live performances, and, most important, urge their affiliated stations to hire more musicians. Stressing again the right of musicians to determine the conditions under which they worked, Weber promised to pull musicians from networks that fed music to affiliates without staff orchestras. Broadcasters, he warned, must come "to the realization that they can no longer have free music unless there is cooperation to employ more musicians."[15]

The networks responded that they had no control over the hiring policies of their affiliates. "You are asking us to do something which we haven't the power to do," CBS attorney Sydney Kaye said. Mark Woods, vice president of NBC, agreed. The networks were "ready to deal for those stations that we own, manage and actually operate," Woods said, but musicians would have to handle their problems with the affiliates through separate negotiations. NAB president James J. Baldwin, with the antitrust laws obviously in mind, accused Weber of trying to maneuver the networks into an unfair labor practice by forcing them to "coerce" and "compel" their affiliates to hire musicians they did not need and perhaps could not afford. But Weber stood firm. "I say to you now," he told broadcasters, "if you insist that the Federation must take our problem up with every individual radio station, it will bring us nowhere."[16]

In a final meeting with industry representatives on August 3, Weber reiterated the union's resolve. "We are willing to make the sacrifice," he said, "and to forgo the employment that we now have in the making of electrical transcriptions or recordings, or even such as we have in the radio industry." Nonetheless, when industry leaders asked for more time to study his proposals, Weber agreed not to strike before September 16. The networks promised to confer with their affiliates and "get down to some basic facts," and the union agreed to form a three-man committee to answer questions the broadcasters might raise. The executive board also agreed to state exactly how many new positions it expected radio to create.[17]

These events marked a turning point in the history of the AFM. Never before had the union reacted so strongly to issues raised by the course of industrial development. Weber had dropped his idea that rallying public support was the only way to save live music, and he said nothing more about the futility of opposing technological change. Advances in technology, he now conceded, were advantageous only if musicians had a meaningful voice in managing—and allocating the rewards of—their consequences. "An employer never concedes anything to a wage worker," Weber told industry leaders, "unless [the worker] uses his economic power." This uncharacteristic bluntness indicated that Weber believed the time for his accustomed accommodation had ended.[18]

Weber's words also showed that musicians had forged a labor ideology of their own. That ideology did not reject capitalism, free enterprise, or economic individualism, nor did it challenge the Constitution, federalism, or the two-party system of government. What it confronted and rejected was the notion that labor was subordinate to capital, or laborers to capitalists, and that laborers had no legitimate say in matters of production and the allocation of profits from production. The ideology rejected as well the argument that labor's collective interests and purposes were threats to the public good. To preserve any sense of independence, equality, and purpose in life, Weber believed, workers had to depend upon their collective power, which was not merely economic and material but moral and ethical as well, and they had an obligation to the public good as well as to their own private gain. Weber had not always articulated such views, but he drew on them now to clarify union objectives and justify resistance.

In mid-August the union offensive began to bear fruit. The first sign of progress appeared when manufacturers of phonograph records—as distinct from electrical transcribers—agreed in principle to a plan put forward by RCA-Victor. That plan was to keep property rights in recordings in the hands of manufacturers but stipulate that their sale was for home use only. At Weber's suggestion the stipulation would be spelled out in licenses manufacturers obtained from the union. Network affiliates, which relied heavily on phonograph recordings, opposed the plan, maintaining that the licensing system put radio at the mercy of the musicians' union and thus violated the Sherman Act. Uncertainty over the legality of the plan in fact forced the union to postpone its implementation.[19]

While network affiliates resisted the licensing plan, they reluctantly accepted other union demands. The networks had made it clear they wanted to avoid a strike, and they urged their affiliates to find ways to appease the union. In a move that reflected the rising tensions between large and small businesses in the industry, some affiliates resisted network efforts to influence their labor policies. A group of about thirty midwestern broadcasters organized to "prevent the networks from herding their affiliates into the AFM fold." However, most large, prosperous affiliates favored conciliation and sought ways to stabilize industrial relations. As a result, a coalition representing roughly half of the affiliated stations organized a new association, the Independent Radio Network Affiliates (IRNA), within the framework of the NAB for the specific purpose of handling negotiations with the AFM.[20]

Despite widespread opposition to the AFM, especially among broadcasters in the South, IRNA agreed to union demands. By the strike deadline its members had agreed to spend an additional $1.5 million annually for live music, double the amount previously spent. In addition, individual stations agreed to pay local union musicians a minimum of 5.5 percent of their gross income from time sales. Industry and union leaders signed a two-year agreement detailing these terms on October 11, 1937, after attorneys had modified the language to avoid conflict with the Sherman Act.[21]

The agreement represented a major victory for musicians. By May 1938 more than 250 affiliates had signed contracts with AFM locals, and radio work for musicians had increased notably. A total of 131 affiliates without staff orchestras before the agreement now had such orchestras, and 77 others had either augmented their orchestras or raised musicians' wages. The musicians' gains, of course, came at radio's expense. In a few cases labor costs skyrocketed. One large station that had spent no money for live musicians in 1937 spent $40,000 in 1938, and another that had spent $6,000 now spent $42,000. The 1937 agreement affected all but a handful of affiliated stations, those that already spent 5.5 percent of their income on live music and those whose gross income was below $20,000.[22]

The agreement with affiliates helped pave the way for settlement with the networks. Although the NAB initially resisted union demands, both NBC and CBS agreed to spend an additional $60,000 annually for musicians at each of their network stations, and Mutual agreed to boost its total expenses for live music by $45,000. The agreement with the networks meant an additional $0.5 million in annual income for musicians.[23]

The employment of instrumentalists in the 386 stations not affiliated with any network had been a low priority for union leaders. The reason for this was the low income of the stations. At the time of the negotiations just concluded, nonaffiliated stations accounted for only a seventh of radio's gross income, and nearly a third of the stations were outside the jurisdiction of AFM locals. Nonetheless, after three days of negotiation in the spring of 1938, the AFM signed an agreement with a newly formed organization representing about sixty of the largest nonaffiliates. This agreement required nonaffiliated stations with gross income in excess of $25,000 (about 145 stations) to spend 5.5 percent of their income on live music, which generally meant no more than a few part-time jobs.[24]

Efforts to regulate the sale of recordings were less successful. Throughout 1938 the union negotiated with manufacturers of phonograph records and electrical transcriptions, making a variety of proposals to control the commercial use of recordings. The problem, as union lawyers had warned all along, was that every one of the proposals might violate the antitrust laws. But that was not the only obstacle. The union had just agreed that broadcasters could use recordings as long as they hired musicians; a boycott of record and transcription companies would jeopardize that agreement. In effect the union had bargained away its leverage in this area. Unable therefore to insist that recorders restrict the sales of records for commercial purposes, the union settled for an arrangement in which manufacturers applied for union licenses that had the legal standing of labor contracts. Licensees agreed to employ union musicians, to maintain specified working conditions, and to secure union approval for rerecording— "dubbing"—over previously recorded performances.

By late 1938, then, the union's militancy had paid off. By threatening to strike, the union had doubled wages in record companies, though Weber insisted that even the new wages were "not in fair relationship to the use made of musical recordings."[25] For single recording sessions of two consecutive hours, of no more than forty minutes' playing time each, musicians now received $24 plus $6 for each additional half-hour of work, while leaders received at least twice those amounts.[26] In addition, the union had curtailed some of the recording industry's most abusive management practices. The AFM was even more successful in the field of radio, where it forged new patterns of labor relations by forcing employers to pay a specified percentage of revenues to workers, one result of which was a significant rise in employment.

The result showed that sound reproduction had not stripped the union of its usefulness, for industry still relied heavily on live musical performances. The best orchestras, like the best comedy and dramatic talent, still made network programming far superior to local programming, which increased the competitive advantage of the networks. In other words, the new agreements helped the networks maintain, perhaps even increase, their industrial dominance. Their affiliates were unable or unwilling to incur higher labor costs on their own, which made them dependent on the networks and thus willing to conciliate the union. The union campaign thus showed that all sectors of radio and recording were susceptible to union pressure and willing to deal with the union to avoid a strike.

The AFM victories coincided with and were not unrelated to parallel advances made by workers in other industries at the time. Labor gains in the late 1930s came chiefly from the new solidarity and militancy of workers, manifested in organizing drives, sit-down strikes, and other forms of activism. A fundamental shift in the attitude of the federal government toward organized labor, which the new activism both helped produce and benefited from, also contributed to the gains. To rejuvenate the flagging economy and provide badly needed relief to millions of hard-pressed Americans, Congress had passed legislation that improved the position of organized workers vis-à-vis their employers. Unemployment compensation, low-income public housing, and federal welfare assistance were among other New Deal policies that benefited the working class.

Of all the new legislation, the National Labor Relations Act, popularly known as the Wagner Act, was of most benefit to organized labor. Guaranteeing workers the right to organize and bargain collectively, the act made possible a dramatic increase in the number of unionized workers. New unions emerged in the steel, rubber, automobile, and other mass-production industries long ignored by the craft-oriented AFL. The Congress of Industrial Organization (CIO), an umbrella organization of the new unions of unskilled and semiskilled workers, was generally more militant than the AFL. By the end of 1937 the CIO had nearly four million members and was a major force in American economic and social life. Embracing militant strategies, CIO affiliates helped millions of industrial workers make meaningful gains in wages, working conditions, and workplace control. This rise of mass activism invigorated conservative unions in the AFL, which saw the CIO as a threat to their own influence in industrial relations. Undertaking new organizing drives of their own, AFL unions increased their membership in the late 1930s from 3.5 million to nearly 4 million. Machinists, carpenters, teamsters, and other AFL affiliates embraced the militant tactics of CIO unions to improve their positions in their respective industries.

This activism had implications for musicians and their union. In 1936 and 1937 the CIO made several attempts to organize instrumentalists and even chartered a few local unions. This created anxiety in some AFM locals. San Francisco Local 6, for example, reported considerable CIO activity among musicians in the Bay Area and urged that the federation "try through every possible means to bring [the] contending parties together." At the 1937 national convention Weber promised to spare "no effort and no expense" in jurisdictional battles with the CIO, and he warned John L. Lewis of the CIO not to "trespass" on the territory of "bona fide" unions. Lewis, who probably thought of Weber as another timid leader of a complacent craft union, made no concessions to Weber's warning, but he never seriously challenged the AFM. The threat from the CIO nevertheless encouraged AFM leaders to take more aggressive action in behalf of musicians.[27]

Thus, the gains musicians made in radio and recording mirrored the general pattern of changing industrial relations. The upsurge of labor militancy and the response of the federal government encouraged employers in radio and recording as well as in other industries to mute their traditional resistance to unions. Industrial conflict in radio and recording no less than in coal or transport promised not only to disrupt profits but also to attract the attention of a federal government now relatively sympathetic to organized labor and relatively suspicious of recalcitrant employers. In fact, other labor problems reflecting national changes in industrial relations diverted the attention of many employers from the problems they had with musicians. NBC's parent corporation, RCA, for example, had to deal simultaneously with the newly organized United Electrical Workers, a far more threatening union than the AFM. In short, industry leaders accepted the National Plan of Settlement partly because it promised to stabilize their relations with musicians while they dealt with what they must have felt were more serious challenges.

THE UNION turned next to the film industry. Since the late 1920s rank-and-file musicians had urged Weber to challenge Hollywood because of the loss of jobs caused by sound movies. Many musicians believed that a union boycott of motion-picture studios could bring back jobs in theaters. Why could not large, profitable theaters feature both live orchestras and sound movies?

The year 1937 seemed a propitious time to confront this issue. The national economy had improved and the movie industry was prosperous. Between 1933 and 1937, unemployment had dropped from 25 percent to 14 percent of the nation's nonfarm workforce, and the real income of Americans had nearly doubled since the depths of the depression five years earlier. Hollywood itself had become a new national symbol of affluence. This was indeed Hollywood's Golden Age, the era of Clark Gable, Shirley Temple, and Fred Astaire, and film had become one the nation's leading industries in terms of assets and volume of business. The major studios were in fact thriving. Since 1935 the net profits of Paramount had risen by more than 800 percent, those of Warner Bros. had risen by almost as much, and those of the leader, Loew (MGM), had doubled. The studios had also increased their control of the exhibition sector of the industry. The seven major studios now owned about fifteen hundred of the nation's seventeen thousand theaters, among them many of the largest and most profitable in the country.[28]

Weber did not call for meetings with studio representatives until 1938, by which time the national economy had turned downward, as had earnings in film. The depth and length of this "Roosevelt recession," however, were not clear when Weber and the union board sat down in New York with the Studio Producers' Committee, a group representing the major studios. At the meeting Weber told producers that before the advent of sound movies, twenty-one thousand musicians had earned nearly $50 million a year in theaters. Now theater work had nearly vanished, and the earnings of musicians in film totaled less than $8 million. The coming of sound movies had thus been "devastating" to musicians, a disaster "incomparable with the destroying of employment opportunities of other workers." Unlike other victims of technological displacement, Weber continued, musicians made the very products that put them out of work: "We are the only craft that makes the medicine that is destroying us." What would happen, Weber wondered aloud, if musicians quit making that medicine? Not wanting a strike, he offered the producers a proposal. If the film studios would hire musicians in their own theaters, the union would consider a reduction in the wage scale of studio musicians in Hollywood. Weber believed that the proposal, if accepted, could generate as many as five thousand new jobs and thus quiet the "clamoring" of the unemployed members of his union.[29]

The Producers' Committee quickly and unanimously rejected Weber's proposal. "For the first time in thirty years," Nicholas Schenck of MGM told Weber, his union had made an "impossible proposition." Theaters got rid of musicians because retaining them had become uneconomic; and it was uneconomic, he reminded Weber, because the public came to theaters to see movies, not to hear musical concerts. Leo B. Spitz of RKO agreed and pointed to recent experiments that had shown the financial impracticality of reemploying theater musicians. To underscore the latter point, the representatives from RKO and Warner Bros. discussed the financial straits those studios then faced, while Albert Warner noted that his company had lost $1.4 million in the last quarter. Paramount's Austin Keough added another reason for opposing Weber's proposal. The Justice Department, he reminded both sides, was suing the studios for violating the antitrust laws by conspiring to restrain trade; and if courts upheld this suit, the studios would have to divest themselves of the theaters they owned. The chances of that happening, Schenck thought, were four to one, and without the theaters the studios would be in no position to help musicians in the way Weber was proposing.[30]

The producers' objections were well founded. The public liked sound movies; if it also wanted live orchestras in movie theaters, that demand would have manifested itself in a willingness to pay higher admission prices, and musicians would be in the theaters. But that demand had not manifested itself; on the contrary, the consumer preference for sound movies over musical performances was everywhere clear. Equally clear was the industry's declining income. In 1938 the net profits of all the studios fell precipitously, with drops ranging from 15 percent in the case of Warner Bros. to nearly 100 percent at RKO. And producers were right about the antitrust suit; it promised to restructure the movie industry in ways that would prohibit producers from hiring theater musicians.

This antitrust suit had its own history, but its implications for musicians were especially pointed. In 1937, when the economy turned downward, Washington had taken new steps to stimulate production and consumer spending. One of those steps involved revitalizing the Antitrust Division of the Justice Department. In early 1938 Roosevelt appointed forty-seven-year-old Thurman Arnold, a hardworking Yale law professor, to head the division, and with increased funding Arnold revamped the division and launched a new crusade to increase industrial and business competition. In May, Arnold began appealing to businessmen and others to notify him of industrial practices that restrained production or competition, promising to prosecute anyone and everyone guilty of such practices. The film industry was one of the first targets of Arnold's campaign. In July he filed twenty-eight separate charges of monopolistic practices against the studios and proposed, among other things, that the courts divest the studios of the theaters they owned.[31]

By 1939 Arnold's antitrust suit had compromised the leverage Weber was seeking to use against the industry. Nonetheless, the suit did not prevent Weber from asking for what he had probably wanted all along: a tariff on the exhibition of films. If the studios could not create theater jobs as Weber had proposed, perhaps the union could "tax" their films and spend the revenues on unemployed musicians. Specifically, the union proposed to levy a "nominal" fee on the showing of "each reel of a picture," a fee that would vary according to theater size but would be so small that it "would hurt no one." In deluxe houses the price of showing a film might rise $40 a week, but in small houses the price increase would be "negligible." Weber believed that the plan would raise between $18 million and $25 million a year, which the union would use to hire musicians to perform free public concerts.[32]

The proposal was unique, radical, or bizarre, depending upon one's interests and perspective. Weber had asked studio management to recognize that technological changes in the methods of production had proved tragic for a specific labor group and to acknowledge that management had a moral obligation to help alleviate the "misery" and "distress" of members of that group. Weber asked specifically for a compensatory unemployment fund, which could provide pride as well as jobs and income for victims of technological displacement. In 1939 no group of displaced workers had ever had such a fund.

To members of the Producers' Committee, Weber's proposal amounted to a privatizing of welfare. Schenck called the objectives of the proposal a "dole" and said that MGM "would have nothing to do with it." The studio, he suggested, would sever its ties with the union rather than agree to such an arrangement. This proved to be the general consensus, and in January 1939 the head of the Producers' Committee, Pat Casey, formally rejected the union proposal. "We cannot accept the financial burden or moral responsibility for the unemployed," Casey told Weber. Producers felt "sympathy" for unemployed musicians but could "do nothing" to help them. The logic of the union proposal, Casey said, would "drive us into a position where . . . the industry must assume the responsibility of furnishing [the worker] with continuous employment or, in the alternative, with relief when his type of service is no longer required. . . . Today the musicians demand relief, tomorrow the former stage hands may demand relief, and the following day the pantomime artists." Casey also denied that the studios had "profited hugely" from sound movies, as Weber contended. To the contrary, they had spent fortunes to survive the transition to sound. "With the introduction of sound," he said, "producers were forced into large capital expenditures, in equipping studios with electrical sound recording apparatus, constructing and remodeling sound-proof stages, completely changing sets and stage lighting and training personnel in the intricacies of a new art." And owning theaters had been no financial bonanza, either. "Theaters [built to exhibit silent films had] to be remodeled and reconditioned for proper acoustical results."[33]

Casey rejected the idea that technological innovation had destroyed more jobs than it had created. "Technological advance did not result in unemployment," he wrote. "On the contrary, what happened was that a *shift* in employment took place and countless thousands of new jobs were created for engineers, electricians, technicians, service men and innumerable others." Producers needed thousands of new workers in sound booths and elsewhere to operate recording equipment, and perhaps even more in the field to exhibit and service the fruits of technological advances. There was no denying that musicians had lost jobs, but those losses were the "cold facts" of technological progress. "It is indeed unfortunate that during the past ten years the displaced musicians have been unable to secure employment in their chosen field. . . . The dislocation of musicians, [however], is no more tragic than the dislocation of skilled workers or artisans which has taken place in other industries as the result of the unrelenting march of the machine age."[34]

Producers thus rejected the union's ideology as well as its demands. To them the sound revolution was an entrepreneurial as well as a financial success, an accomplishment not of predatory businessmen using labor-saving machinery to displace workers but of men of social vision as well as business acumen. Their obligation in its aftermath was not to musicians narrowly defined as workers but to the general public as beneficiaries and consumers of the new forms of entertainment and leisure activity that the sound revolution brought to the masses. Among those beneficiaries were many musicians themselves, in those hundreds of high-paying steady jobs in Hollywood studios. "We have done all we can possibly do," Casey told the union representatives.[35]

Weber of course rejected these views, but he responded to them by demanding more bargaining sessions, which he postponed twice, the last time indefinitely. There were good reasons for this restraint. By 1939 the seriousness of the economic downturn had become clear. Nonfarm unemployment had climbed back to 19 percent in 1938, and it was still 17 percent in 1939. The fortunes of the studios paralleled these developments. After a nearly disastrous year in 1937–38, profits began to rise again, but the studios were still losing money in 1939, especially Warner Bros., Universal, and RKO. Meanwhile, the government's antitrust suit was proving to be especially distracting; when it finally came to trial in 1940, it had been postponed thirteen times. A union "tax" on films would have strengthened Arnold's argument that the major studios had monopolized the supply of films, and it might have inadvertently drawn Arnold's attention to the AFM's own insistence on minimum-size orchestras in radio.[36]

Then, too, it was not clear that the union could win a studio strike. That the industry depended upon fewer than five hundred musicians was not a fact the union could necessarily use effectively. To ask a relatively few high-paid musicians to strike for someone else's benefit was not a sure formula for success, especially after Weber had told the producers he would consider a reduction in those same musicians' pay in return for additional jobs in theaters. Whether Weber's failing health was an additional factor in the handling of the strike question is unclear, but it made the union's immediate future uncertain.

THERE WERE other considerations as well. The National Plan had created unexpected problems. Recording musicians found that higher wages resulted in less work. Faced with rising wages and stagnant sales, producers responded by recording smaller bands and orchestras. A single organist might replace as many as six instrumentalists, and popular bandleaders began practicing what a later generation called downsizing.[37]

A revitalization of the NAB also pointed to future problems. Many broadcasters believed that acceptance of the National Plan reflected a failure of NAB leadership and the association's inability to protect the interests of radio. Two executives of WBEN in Buffalo aired these views in a letter to hundreds of broadcasters. "The condition will become worse rather than better," they warned, "unless we take ourselves by the bootstraps and apply effective, corrective measures." In response to this new call to arms, broadcasters reorganized the NAB in late 1937, creating a separate department to deal exclusively with labor problems and increasing membership dues by a combined total of $80,000 to $120,000 a year. (The voice of network affiliates, IRNA, remained, at least temporarily, a separate entity within the NAB.)[38]

Criticism of the National Plan mounted as the expiration date of the plan approached. Network affiliates complained that the plan was "an onerous burden," while broadcasters in small cities complained that they were unable to "sell" local staff orchestras to advertisers. The quality of local musicians, they argued, was inferior; audiences and advertisers alike preferred recorded music and network programs. In fact, many broadcasters, especially in the South and West, had hired musicians as the agreement dictated but refused to use them. These broadcasters understandably complained that the National Plan was "an enforced payment of tribute," a "subsidy" exacted from them by the union. *Broadcasting*, the trade journal of the industry, gave voice to the growing discontent. The agreement with musicians, one of its editorialists complained, was a "miserable flop . . . procured under duress," while another complained that broadcasters had signed the agreement "with a sword hanging over [their] heads."[39]

Local branches of the AFM, however, were insisting that the provisions of the agreement be expanded to include more musicians. This sentiment was strong enough to cause Weber, in November 1939, to ask radio to boost the size of staff orchestras. In negotiations conducted not long before the agreement was to expire, Weber proposed that the networks agree to double the money they paid for staff musicians and that their affiliates spend an additional $1.5 million. Invoking the tactic that had worked two years earlier, Weber threatened a general strike if radio failed to accept these demands.[40]

Stage to Studio

Weber justified the demands on the grounds that industry profits had soared under the agreement. Indeed, radio's experience had differed from that of film. Net profits rose dramatically in the several years around 1940. Typical headlines in *Broadcasting* read, "All Time Sales Records" and "Radio's Best Year." Gross time sales of broadcasters in 1939 were 14 percent higher than in 1938, and in 1941 they were 21.5 percent higher than in 1939. This steady and substantial rise reflected the growing numbers of radios in American households and consequent increases in advertising revenue. By 1940 more than 80 percent of all homes in the nation had radios, and Weber thought musicians deserved a fair share of the rising wealth.[41]

Broadcasters disagreed, or rather they had a different view of what constituted a fair share. They denounced Weber as "dictatorial" and the AFM as a "parasite" whose thirst for blood would end up "killing the goose." Repeating those views, network affiliates "flooded" IRNA with "letters, telegrams and phone calls," insisting, in the words of Samuel R. Rosenbaum, who headed the new organization, that IRNA reject Weber's "outrageous" and "unfair" demands. On November 20, 1939, Rosenbaum announced that the affiliates would not renew the National Plan, proposing instead that the union negotiate agreements with each affiliate "without reference to any national plan or quota."[42]

In taking this stance the affiliates had no doubt been encouraged by recent actions of the Justice Department's Antitrust Division. Only one day before Rosenbaum notified musicians that the National Plan of Settlement would not be renewed, Thurman Arnold sent a letter to the Central Labor Union in Indianapolis detailing practices in the building trades that he considered "unreasonable restraints of trade" and "unquestionable violations" of the Sherman Act. Among the practices he listed were those designed to compel the hiring of "useless" or "unnecessary" workers. Broadcasters recognized at once the potential applicability of these descriptions to provisions in the National Plan. So too did the Justice Department. Less than two weeks after Rosenbaum rejected Weber's proposals, the Federal Communications Commission (FCC) informed him that the Justice Department would act against the musicians as soon as broadcasters filed a formal complaint with the agency. Rosenbaum did so at once, citing the "enforced employment of musicians" as a restraint of trade upon radio.[43]

News of the resulting antitrust suit was soon reverberating through AFM headquarters. Union leaders feared that the courts would uphold Arnold's definition of restraint of trade, and thereby largely destroy musical employment in radio outside media centers. The union therefore made a strategic retreat. On January 17, 1940, Weber announced that he would not seek renewal of the National Plan, and he instructed union locals to "use their own judgement in entering into contracts with the radio stations in their jurisdictions." He also withdrew the threat of a nationwide strike.[44]

Broadcasters hailed this as a "major victory" for free enterprise. It was no victory, however, for working musicians. When the 1937 agreement expired in February 1940, network affiliates were free to hire musicians or not as they saw fit. Again the availability of recorded music and network programming had undercut the union's bargaining power. But if the collapse of the National Plan eliminated the requirement to hire musicians, it did not cause the musicians to leave their jobs quietly. On the contrary, from Florida to Minnesota, musicians did whatever they could to protect their position in radio. James C. Petrillo, who had recently succeeded Weber as president of the AFM, coordinated their efforts.

PETRILLO HAD WON the union presidency by unanimous vote at the annual convention in Indianapolis in 1940. His election, which came after Weber announced his retirement for reasons of health, had been long anticipated. Musicians across the country, especially those in small locals and those devastated by the impact of technological innovation, considered Petrillo, head of Chicago Local 10, as the man most likely to do something effective about their problems. The Chicago union was not only militant; it had also pioneered ways of responding effectively to the problems caused by recorded music. In 1935, to illustrate, Local 10 had pressured Chicago broadcasters into agreeing to destroy records after the stations had used them once, and even before the National Plan was in place the local had prohibited its members from recording for manufacturers it had not licensed. The licenses not only protected wages and working conditions in record companies but also restricted record sales to commercial companies deemed fair to labor. This restriction, which lasted a year, cost Chicago musicians perhaps $150,000 in lost wages and prompted several recording companies to move out of the city. The union action, however, made it clear that Petrillo would do whatever he could to protect musicians from the impact of recorded music.[45]

Petrillo's willingness to fight sprang from no naive belief that workers could "uninvent" technology. He recognized that unions could not, any more than management, hold back technological change. "From the time when organized labor first engaged in warfare against the use of machinery," he said in 1928, "conflict has resulted in the complete, or partial, destruction of the union involved." Petrillo believed, however, that musicians were in a position to influence the impact of recording technology on their work and thus on their well-being. "In the endless conflict between labor and machinery," he stated, "the musician is more favorably situated than any other worker." Unlike workers in mass-production industries whose skills new technology made marginal, "the living musician," Petrillo said, "must be consulted and his services utilized . . . or else the machine will be silent." The Chicago leader hoped to control the use of sound technology by enlisting the support of instrumentalists who made recorded music.[46]

Petrillo's background partly explains his tough-minded approach. Son of an immigrant Italian father who supported his family by digging sewers, Petrillo grew up in a rough, run-down neighborhood on Chicago's West Side. He had little education—he never finished elementary school—and even less refinement or tact. Nor was he, by his own admission, a good musician (he played the trumpet). What he was was a union man. In 1914, at the age of twenty-two, he was elected by his fellow musicians to the office of president of the Chicago chapter of the American Musicians' Union, and when they denied him reelection three years later, he switched his allegiance to the Chicago local of the AFM. A year after that the four thousand members of the local elected him vice president and, in 1922, president. For eighteen years the diminutive man—he was five feet four inches tall—oversaw an organization many of whose members were variously associated with gangsters and with the buying and selling of alcohol during the Prohibition era. It is not surprising that he rode around Al Capone's Chicago in a bulletproof car and walked about it with armed bodyguards. In 1924 a bomb exploded on the porch of his home.[47]

By 1940 broadcasters and recorders were well aware of Petrillo's reputation for confrontation. A member of the AFM executive board since 1932, Petrillo played an active role in the negotiations that resulted in the National Plan. Commenting in 1940 on his elevation to the AFM presidency, *Broadcasting* predicted a "new and more vigorous regime for union musicians," one that would likely produce "increased demands upon radio." Petrillo, the trade journal noted, had pledged a "fight to the finish" against Thurman Arnold's effort to limit the power of musicians to protect their jobs in and out of radio. Broadcasters expected Petrillo to launch a campaign against radio, for he had long argued that the union should exercise greater control over musicians who worked for network stations.[48]

Petrillo did indeed have plans to shake up radio. Specifically, he planned to help musicians by restricting remote broadcasting. Radio performances of traveling big bands had become increasingly popular in the 1930s. During prime-time hours at the end of the decade, the networks regularly broadcast to their affiliates the live concerts of Benny Goodman, Artie Shaw, and others. Petrillo hoped that by occasionally forbidding musicians to make remotes, he could force the networks and their affiliates to accept higher wages and minimum orchestra sizes. Although remotes were less important than staff orchestras for network programming purposes, Petrillo was willing to gamble that the networks would be reluctant to do without such high-profile programs. If pulling the remotes proved ineffectual, he was prepared to pull staff orchestras from network-owned stations. The plan, of course, demanded union solidarity.

Saxophonist Henry Gruen later shed light on the work patterns behind Petrillo's maneuverings. In 1937 and 1938 Gruen traveled with Jimmy Joy's twelve-piece orchestra, performing at hotels and dance clubs throughout the Midwest. For two or three weeks at a time the band worked at the Stevens Hotel in Chicago. "We played there six nights a week," he recalled. "We started about eight and played till one. [We] played songs like 'Ain't She Sweet,' 'After You've Gone,' and 'Maria.'" The networks sometimes picked up these performances. "We loved it," Gruen said of remote broadcasting. "It helped you make a name for yourself. When I did a solo the announcer would say, 'Alto solo by Henry Gruen.'" The remotes also helped the band get more and better jobs. "They made us popular," Gruen explained. "People would flock in to see us because they heard us on radio." Gruen had been unaware of Petrillo's specific plans to restrict remote broadcasting, but he had liked Petrillo and knew "he planned to quit giving our product away." He reminisced, "We thought he would be great because he did so much for Chicago. He was truly for the sideman. He was one of those guys like Truman [who said], 'The buck stops here.'" Gruen not only trusted Petrillo; he also was prepared to sacrifice to help him accomplish his goals. "I saw [theater owners] cover up the pits with canvas," he recalled of the loss of theater jobs, "and felt something needed to be done [to prevent a similar loss] in radio. So many people were making money off the sweat of musicians. They were bleeding the workingman to death."[49]

Petrillo's charisma and confrontational style would obscure the role of the AFM executive board in industrial relations, yet the board significantly shaped union policies throughout Petrillo's presidency. In 1940 the board embodied a wealth of experience and expertise. Thomas Gamble, a board member since 1908, had served as president of Local 198, Providence, and worked on the legislative committee of the Rhode Island State Federation of Labor. Chauncey Weaver, who joined the board in 1915, had headed Local 75, Des Moines, and worked as a musician, lawyer, and journalist. Charles Bagley, vice president of the AFM since 1931, was one of the union's principal legal advisers and had previously held office as president of Local 47, Los Angeles. Oscar Hild, who joined the board in 1940, had served as president of Local 1, Cincinnati, since 1931.[50]

Disputes between network affiliates and AFM locals quickly provided occasions for Petrillo and the executive board to demonstrate both leadership ability and union solidarity. In June 1940 officials at the St. Paul local asked Petrillo for assistance when negotiations with station KTSP broke down. Since KTSP was an NBC affiliate, Petrillo notified the network that unless the dispute was settled within twenty-four hours to the satisfaction of the local union, he would prohibit traveling bands from performing for NBC remote broadcasts. When that condition was not met, Petrillo made good his threat. On June 28 he ordered ten big bands to cease playing on NBC. Among the bands affected were those of Woody Herman, Tommy Dorsey, and Gene Krupa. When KTSP still refused to meet his demands, Petrillo promised to pull staff musicians from NBC itself within twenty-four hours. The following day KTSP and the musicians settled their differences in a manner Petrillo described as "satisfactory to the St. Paul Local."[51]

While these events transpired, a similar dispute arose in Richmond. There the struggle involved musicians at WRVA, an affiliate of CBS and MBS, who requested Petrillo's help when the station proposed to lower their wages. After investigating the situation Petrillo gave Richmond the same assistance he gave St. Paul. He ordered eight traveling bands to discontinue remote broadcasts to CBS and Mutual, and over the next few days he tightened the noose by ordering twenty-two others to do likewise. On July 10, after Petrillo had pulled a total of thirty scheduled performances from the two networks, WRVA abandoned its plan to lower the wages of its musicians.[52]

Petrillo had protected wages and working conditions in local areas by withdrawing programming from the networks. His gamble had paid off. The St. Paul and Richmond episodes had shown what Petrillo suspected: that affiliate stations were willing to resist the union, but the networks were not. The affiliates had used network programming to undermine union locals, and the networks now used the union threat to that programming to force them to agree to union demands. The affiliates, in other words, could not prosper, perhaps could not even survive, without network programming. The movie and radio industries had long relied upon the star system; now, cleverly, the musicians' union was putting that system to its own use.

Petrillo was determined to lose no radio staff orchestras. In mid-October 1940 he told representatives of the three networks that beginning January 1, 1941, he would permit no remote performances that created "unfair competition" for local musicians. In January, accordingly, he canceled remote performances to protect jobs in Akron and Scranton, and in February he did the same thing for Nashville. To the ire of broadcasters and audiences alike, the music of Lawrence Welk, Duke Ellington, and Artie Shaw become unavailable in those cities. Petrillo's tactic was working. "I believe that the chains are beginning to realize," he said in July 1941, "that when I say a strike will be called, it will be called."[53]

Musicians in traveling bands made few complaints when Petrillo prohibited remote broadcasting of their performances, even though they sometimes lost income as a result of the prohibition. Petrillo banked on the solidarity of the musicians, and he appreciated their sacrifice. "When remote control bands from coast to coast were directed to discontinue services they did so at once," he noted in 1941, "without even one of the bands involved questioning what it was all about." Petrillo also acknowledged the role of local union officials in the success of his effort. "I want to thank the officers in Locals involved," he said also in 1941, "for the splendid cooperation given my office by each and every one of them." This cooperativeness was partly a tribute to Petrillo's persuasiveness, but it was also due to big-band musicians' recognition of the strength of the AFM. Their own jobs, in other words, depended on their cooperativeness. Their cooperation, as it turned out, cost them little. Petrillo may not have wanted remote broadcasting at all, but he realized that since it existed it was to his advantage to manipulate it rather than attempt to destroy it.[54]

WHILE PETRILLO struggled to protect radio staff orchestras, the problem of canned music persisted. Jukeboxes continued to displace musicians by the thousands. In 1941–42 six thousand jukebox operators had perhaps four hundred thousand of the machines in operation, which by Petrillo's estimate deprived musicians of eight thousand jobs. That estimate may have been high, but many clubs and dance halls that had once featured live music now operated as "juke joints."[55]

"Hepcats" and a jukebox, circa 1940. Nickel-in-the-slot machines created hit tunes and popularized orchestras; they also displaced live musicians—prompting James Caesar Petrillo, then president of the American Federation of Musicians, to launch a national campaign to regulate the commercial use of canned music. *(Bettmann Archive)*

Frequency-modulation (FM) broadcasting and television were other recent innovations with major implications for the future of working musicians. FM broadcasting was the stepchild of Edwin H. Armstrong, a New York inventor who had made an earlier contribution to radio with the development of feedback circuits. In 1933 Armstrong first demonstrated FM, which transmitted sound at higher wavelengths and within a wider band of frequencies than the traditional amplitude modulation (AM). As a result, FM transmissions were virtually static-free. Although FM receivers did not become generally available for several years, broadcasters demonstrated FM technology to civic and business groups across the nation, and interest in the new process was high. By 1941 mass production of FM sets was under way, spearheaded by General Electric and Philco. Stations soon flooded the FCC with applications for FM licenses. Musicians, however, realized that FM would mean new jobs only if the recorded music and remote broadcasts that were the staples of AM radio did not also become the bulk of FM programming.[56]

Meanwhile, the first public display of television took place in 1939 at the New York World's Fair. AFM leaders, however, knew already that the advent of television was imminent, and they knew too of the interest of the radio networks in it. By 1937 NBC and CBS had built television transmitting stations, and Zenith, Philco, General Electric, and Dumont Laboratories were manufacturing television sets. Union officials no less than industry leaders were uncertain of the future of the new medium, but the former hoped and anticipated that the public wanted to see the performances they had heretofore only heard. Petrillo doubted that television would use live musicians unless the union put up a fight. In 1941 he assured members that the union was preparing "the musicians' platform for commercial Television . . . when [television] comes into general use."[57]

There were still other concerns caused by new technology. Wired-music companies had developed an experimental "telephone jukebox," a machine that allowed an individual to drop a coin into a slot, pick up an attached microphone, and request that a certain song be played. At a nearby station a disc jockey—"pancake turners," in union lingo—played the requested recording, its sound transmitted to the requesting machine through electric power lines. Still another development, this one in sound-on-film technology, had resulted in coin-operated machines that projected onto screens filmed scenes of bands playing or musicians otherwise performing to the accompaniment of their recorded music. By 1941 Alvino Rey, Count Basie, and other popular musicians had appeared before these "Soundie" cameras. These "coin-operated movies" were still "an unknown quality" to Petrillo, for they were not yet commercially practical. But a few of the machines were around, and musicians and the union eyed them warily.[58]

Local branches were fully aware of these threats, and delegates to the 1941 national convention in Seattle offered a score of proposals concerning them. Those from Local 453 in Winona, Minnesota, urged the executive board to get the 1909 copyright law amended to give musicians ownership of the music they recorded in the same way authors owned the lyrics and scores they wrote, or indeed the books they published. Other locals suggested that the board concern itself with closer regulation of the distribution of records, including restricting the sale of records to home use only. Local 616 of Salinas, California, offered an elaborate plan to require juke-box operators to pay the union 65 cents for every 35-cent record they made available to the public, and 50 cents for each of the 50-cent longer-playing records many of them preferred, in effect making each record cost operators $1. The "label fund" thus generated would be distributed according to a formula under which the local in whose jurisdiction the jukebox operated would receive 78 percent. The local would use the money to police the jukebox business and help unemployed musicians. The remaining 22 percent of the fund would be divided between the national union and the musicians who had made the records.[59]

While union leaders weighed these proposals, new labor disputes dramatized the circumstances that had generated them. Until June 1, 1941, to illustrate, the workforce of Ringling Brothers' Circus included forty instrumentalists: twenty-six white musicians who played for main events at $47.50 a week each, and fourteen black musicians who worked the side-shows at rates of $26.50. When the circus management refused to increase each of these rates $2.50, the musicians went on strike, during which time management used recorded music and the show went on. The strike ended when Petrillo and Robert Ringling agreed, for reasons that are unclear, to a wage settlement that exceeded the original demand. Nonetheless, the dispute showed again the reality of the threat posed by recorded music.[60]

The threat surfaced again in radio. In October 1941 H. J. Brenner, who owned CBS affiliate WJAS in Pittsburgh, refused to renew a contract requiring him to employ an eleven-piece orchestra. When negotiations deadlocked, musicians at the station went on strike, as did those at station KQV, the NBC affiliate in Pittsburgh, which Brenner also owned. Since the strikes involved the two leading networks, it had national implications. Petrillo's response to the first strike was to forbid musicians to provide remote services to CBS. Then, in a preemptive move that raised the stakes, NBC canceled its own upcoming remotes, replacing them with talk shows and staff orchestra concerts and simultaneously transmitting popular programs over its Red and Blue channels.

These moves did little to help the networks. Canceling the performances of bandleaders like Benny Goodman, Gene Krupa, and Glenn Miller while the union withheld similar bands from CBS gave Mutual a monopoly on the music of traveling bands and a lion's share of late-night audiences. Moreover, the general policies of the networks concerning musicians convinced Petrillo that the time had come for a showdown. Accordingly, on October 14 he ordered musicians on selected NBC and CBS sustaining programs to cease work, promising to reimburse them for the wages they lost. The order affected musicians in NBC studios in New York, Chicago, Hollywood, San Francisco, Detroit, and Cleveland, as well as those in CBS studios in New York and Chicago. It did not affect commercial programs, however, though Petrillo threatened to include musicians on those programs unless Brenner settled quickly with the Pittsburgh musicians.[61]

Brenner did so, and on October 17 Petrillo accordingly lifted the restrictions on remote and sustaining broadcasts on the two networks, and the musicians at WJAS and KQV went back to work. Although the Pittsburgh local agreed to minor concessions, Petrillo and the AFM had won an important victory. At a cost of $5,000 Petrillo and the union had successfully protected radio jobs outside the media centers, and he was certain the networks had pressured their affiliates to agree to his demands.[62]

Nevertheless, conflicts with radio continued. By the summer of 1942 Petrillo had intervened in more than twenty local disputes, in each case preserving radio jobs by pulling remote performances. But each time he did so he encountered anew the tenuous position of musicians in radio. The underlying problem was that the union could not afford to pay the wages of striking musicians for an extended period of time. Moreover, once he canceled remote performances, Petrillo had used the only effective weapon he had.

Paradoxically, while radio musicians worried about losing their jobs, other instrumentalists were frantically producing the products that threatened radio employment. In April 1940, for example, members of the Tommy Dorsey band were working in two recording studios a day, from early morning till late at night. Band member Joe Bushkin recalled the hectic pace. "We started at 9:00 or 10:00 in the morning and had a break for lunch, then worked through 6:00 or so. Then we had just enough time to get uptown and grab a sandwich or something. . . . So we showed up at the second studio at about 8:00 in the evening and we played until about 1:00 or 2:00 in the morning." Sidemen like Bushkin received $125 a week for these recording sessions.[63]

Musicians no less than Petrillo understood the irony of these things. At the union's annual convention in Dallas in 1941, delegates urged Petrillo to "draw the line" against the real nemesis, the unrestrained use of recorded music. He promptly did so, announcing to the convention on June 8, 1942, that union musicians would no longer "play at their own funeral." They would, that is, make no more records or electrical transcriptions— ever. Delegates received the announcement with thunderous applause.[64]

Petrillo soon set the ban in motion. He told recording and transcription companies that beginning August 1, 1942, union musicians would "not play or contract for recordings, transcriptions or any other form of mechanical music."[65] To blunt charges that the ban was a secondary boycott or otherwise in restraint of trade, Petrillo told James L. Fly, chairman of the FCC, that he and his union made no demands on industry and had nothing to negotiate with anyone. Instead, he and they had ceased recording because recorded music deprived them of work. Petrillo told Fly, "95% of the music in the United States and Canada is canned music. Only 5% is left for the poor professional musician who studied all his life so that he might make a living for his family. This is not a question of being a 'czar' or 'dictator,' it is a question of a large group of men fighting for their very existence."[66]

PETRILLO'S ANNOUNCEMENT and its subsequent implementation showed that since the difficult years of the 1930s, the AFM had regained considerable power and solidarity. Since 1936, union membership had jumped from 105,000 to 135,000, and though many of the new members lived in New York, Chicago, or Los Angeles, locals around the country had witnessed meaningful gains. Membership in Baltimore, Memphis, and San Francisco, for example, had risen by 10 to 20 percent; growth in Minneapolis and Kansas City was even greater. Many small locals, about eighty or so, resurfaced in these years. The union had not fully recovered from the twin blows of mechanization and the depression, but the rising membership clearly encouraged Petrillo's confrontational strategies. His rhetoric, however, fooled no one. Friend and foe alike knew that what he wanted was jobs for members of his union, and that he hoped somehow to create them by reducing the competition from recorded music. It was not clear that he could do so. Broadcasters had a powerful association committed to defending their interests, and many in the Justice Department and the federal courts still thought of the things Petrillo was proposing in terms of secondary boycotts and restraint of trade. In addition, consumers clearly preferred the canned music of national performers to the live product of most local musicians.

Nonetheless, musicians and their union were intent on doing what they could to protect themselves from the consequences of products of their own making. Like other workers, they rejected the notion that new technology should benefit only consumers and capitalists. Workers too deserved to benefit. It was largely musicians who had made radio and recording what they had become, and profits in these industries were at all-time highs in 1941 and 1942. "Why can't we all live?" an impassioned Petrillo asked on behalf of musicians. "Why should all big revenues go into the pockets of these radio stations . . . without paying one dollar to the musicians?"[67] Why, indeed?

Stage to Studio

Six

Recording Ban

THE BAN ON RECORDING alarmed the music industry. Without fresh supplies of recordings, record companies, jukebox operations, and music retailers faced disaster. Phonograph manufacturers and radio stations dependent on the consumption of popular music were similarly imperiled. What could be done to get recording musicians back to work? Unlike other striking workers demanding higher wages, better working conditions, and greater control of the workplace, the striking musicians said simply that they would make no more of the instruments of their own displacement. Clearly, solutions that worked in other labor-management conflicts would not work in this one.

The story of the recording ban, which lasted from August 1942 until November 1944, is significant in retrospect because of the insights it offers into the links between work, technology, and industrial relations. It shows how a distinctive group of workers tried to control the distinctive product of their labor and with it the deployment of new technologies that threatened their well-being. It also shows that employers fiercely resisted the initiatives of labor with sophisticated campaigns designed to protect their own prerogatives. This contestation reveals too that struggles between labor and management were not always fought out on factory floors or outside company gates. On the contrary, they also took place in courtrooms and in myriad battles for the hearts and minds of the American people, few of whom knew or cared much about record production.

WHEN JAMES C. PETRILLO, president of the American Federation of Musicians (AFM), set in motion the union's "ultimate weapon" against recorded music, industry leaders pledged an "all-out" campaign against his "gangster acts." On June 15, 1942, six weeks before the ban actually began, *Broadcasting* predicted that radio "will not take this one lying down." Petrillo was out to "wreck" radio, the trade journal insisted, and only a united, aggressive stand against him and his union could save the radio and recording industries. The challenge would be formidable, the journal admitted, but it reminded its readers that radio had recently broken the monopoly of another labor organization, the American Society of Composers, Authors, and Publishers (ASCAP).

The victory over ASCAP had indeed been impressive. Since 1923 ASCAP had collected performance fees from broadcasters that used the music it licensed. In the late 1930s, however, the National Association of Broadcasters (NAB) began resisting those fees. By 1941, with strong support from NBC and CBS, employers had effectively compromised ASCAP's bargaining power through their own expansionary strategies of vertical integration. By building a competitor to ASCAP beholden to themselves—Broadcast Music Incorporated (BMI)—they effectively undercut the association. Through BMI they secured rights to music controlled by small competitors of ASCAP, and then favored that music over recordings licensed by ASCAP.[1] The level of cooperation between the networks in this endeavor was impressive, and it undoubtedly encouraged them to think they could repeat the success with the AFM—that is, to think they could stabilize their relations with the musicians they employed on their own terms and without jeopardizing their competitive position in the marketplace.

There were, however, important differences between the two organizations they challenged. Unlike ASCAP, which faced competition from dozens of small organizations of composers and music publishers, all of which had reason to cooperate with radio, the AFM had no competitors. It had instead a monopoly on musical services guaranteed by closed-shop hiring policies throughout the industry. Musicians thus knew that to violate the recording ban would be to deny themselves employment. Trombonist Bill Hitchcock, who recorded with the Eddy Duchin Band in 1940, remembered that he never considered breaking union ranks. "In those days," he recalled, "we would not have done that. We wouldn't strike-break." Reedman Will Brady, who recorded with the Ozzie Nelson Band in New York shortly before the ban began, agreed: "We wouldn't do anything to jeopardize our standing in the union." Both men offered another reason for the solidarity: few musicians depended only on their earnings in recording. "Recording wasn't our whole life," Hitchcock recalled, "it was only the icing on the cake." Or, as Brady put it, "I did a lot of recording, but it was never my main source of income." Unlike other striking workers, then, recording musicians honoring the ban were not risking their bread and butter.[2]

Industry leaders recognized the advantage this gave the union, but they had strengths of their own as well as a calculated strategy. The first step in that strategy became apparent even before Petrillo announced the ban on recording. Anticipating the ban, recorders stockpiled as many new recordings as possible. In the weeks before the ban, instrumentalists found themselves overwhelmed with work, recording with such popular artists as Johnny Mercer, Kay Kyser, and Dinah Shore. Among the last things thus recorded, only hours before the boycott began, was Kyser's rendition of "Praise the Lord and Pass the Ammunition."[3] The stockpiling was a clear sign that industry leaders expected and were preparing for a protracted struggle. Indeed, it almost guaranteed that result if musicians maintained their solidarity and determination. The expense of creating such a huge stock of unreleased recordings made it financially impractical to resume record production immediately.

After the ban began, record companies found additional ways to withstand its effects. Some of them made recordings of singers backed by choral groups or nonunion musicians playing harmonicas, ocarinas, and other instruments not covered by AFM rules. Others reissued old favorites. But these and other innovations were peripheral to a larger strategy. To defeat Petrillo and the union, industry leaders would rely chiefly on public relations and union-busting. On behalf of the networks and some six hundred radio stations, the NAB coordinated what became a classic antilabor campaign. CBS attorney Sidney Kaye, who played a leading role in the creation of BMI, served as NAB counsel in the effort. When the boycott began, Kaye and NAB executives had already met in Chicago to map their strategy.[4]

As employer associations often did in labor conflicts, the NAB relied on the press to help discredit union leaders and their purposes and actions. The association paid the New York public relations firm of Baldwin and Mermey $2 million to "articulate industry's position" and "activate the public" against the union. Newspapers across the nation were soon lashing out at Petrillo's character as well as his words and actions. The *Chicago*

Daily News referred to him as "the inflated little nonentity who strong-armed himself into dictatorial power." The *Washington Post, Buffalo News,* and other newspapers carried equally caustic remarks, variously labeling the union leader "czar," "tyrant," and "dictator of the music realm." Editorialists often pointed out that Petrillo's middle name was Caesar, which *Broadcasting* suggested "was unquestionably given him by people who foresaw his future."[5] Such name-calling no doubt influenced public perceptions of Petrillo, but the editorials in which it appeared diverted attention from the issues the boycott raised. None of the editorials discussed the impact of mechanization on the employment opportunities of musicians or even acknowledged that Petrillo was the democratically elected head of a union with 513 locals, all of which supported the recording ban.

Editorials attacking Petrillo mirrored the mutual interests of employers and the press. Dependent upon revenue from advertisers, the daily press had over the years generally sided with employers in labor disputes. A special problem Petrillo faced was that newspaper interests owned about a quarter of the nation's radio stations, about two hundred of them, and thus had a stake in discrediting the AFM. Petrillo himself placed the number of stations owned by newspapers at nearly three hundred, all of them, he believed, "on the side of the employer and against the federation whether we are right or wrong."[6] Even in exaggerating this situation, Petrillo recognized that the public perception of radio as an entertainment medium obscured the fact that it was also an industrial enterprise with its own economic imperatives, among them the desire to control its workforce.

In responding to the striking musicians, employers effectively exploited longstanding tensions between labor activism and patriotism in a time of war. Throughout the labor boycott, employers accused Petrillo and the boycotters of disloyalty, and the accusation influenced public perceptions of both. The boycott did indeed defy the spirit of commitments representatives of labor and management had made to work cooperatively for the duration of the war. Shortly after Pearl Harbor, leaders of the American Federation of Labor (AFL) and the Congress of Industrial Organization (CIO) had made general pledges not to strike in return for no-lockout clauses in labor contracts and increased government protection of trade union activity. This agreement made labor leaders who supported wartime strikes vulnerable to the kind of criticism Petrillo was now receiving. The NAB argued that Petrillo and his union were not only violating the no-strike agreement but also denying the public a product essential to the war effort. "Music plays a vital part in war morale," the association argued; Petrillo's actions were thus "unpatriotic" as well as "arbitrary and illegal."

Some employers went further than that, likening Petrillo to an enemy leader. Stanley E. Hubbard, general manager of KTSP in St. Paul, called Petrillo a "fuhrer" with a "public be damned attitude." "Petrilloism," Hubbard said, is "as cruel and brutal as Hitlerism."[7]

Employers also tried to undercut Petrillo by characterizing the recording ban as Luddism, a furiously futile effort to wreck the technology of musical progress by what amounted to machine-smashing. One editorial in the employer-controlled press compared "Little Jimmy" with "the short-sighted men who battled the introduction of the spinning jenny" and added that the recording ban was "merely another chapter in the prolonged battle . . . against technological progress." Such statements made opposition to technological innovation appear backward and irrational, even as they linked the advance of technology to the rise of corporate profits. In response to such charges, Petrillo insisted that he opposed not technological innovation but the ways employers used technology to benefit themselves at the expense of their workers.[8]

While maligning Petrillo in the press, the NAB lobbied state and federal lawmakers for legislation against the AFM and the ban. The association's influence at the state level had been apparent in the late 1930s, when North Carolina, South Carolina, Florida, and other states passed bills allowing radio stations to broadcast records with labels that prohibited that use.[9] At the national level, NAB lobbyists endorsed the view of Thurman Arnold and the Justice Department's Antitrust Division that the AFM could not, any more than unions in the building trades or transport, force employers to hire "more employees than needed." In fact, broadcasters worked closely with Arnold to undermine the ban. Only a week into the ban, representatives of the broadcasting and recording industries met with Arnold to discuss ways to curb Petrillo's "onslaughts" through litigation as well as legislation. Employers' hopes rose in the summer of 1942, when Arnold filed a complaint in federal court in Chicago that the AFM, in violation of the Sherman Act, had "unlawfully combined and conspired to restrain interstate trade and commerce in phonograph records, electrical transcriptions, and radio broadcasting." Asking for an injunction to end the ban, Arnold told the court that the ban threatened to destroy entire industries, including record and phonograph manufacturing and retailing, the jukebox industry, and many radio stations.[10]

As the Justice Department prepared to argue in court against "unneeded" labor, a long-simmering dispute between the AFM and the National Music Camp at Interlochen, Michigan, compounded the union's problems. For twelve years this popular summer camp for children had concluded with a musical festival, the highlight of which was always a concert by the camp's amateur instrumentalists. The concert concerned the union because NBC transmitted it to its affiliates. The children, who participated in a national competition to attend the camp, practiced all summer for the chance to be heard on radio.

In accordance with standard union policy, the sponsor of the broadcast, Majestic Radios, originally paid a standby fee to the AFM local in Interlochen for permission to use nonunion musicians within its jurisdiction. That arrangement, however, lapsed in 1931. When Petrillo became president of the AFM, he demanded its reinstatement. Insisting that the National Camp, its sponsors, and NBC were commercial enterprises, he also demanded that they use union musicians whenever they promoted their services or products. When NBC informed Petrillo that it had already made arrangements to broadcast the children's concert in 1941, Petrillo uncharacteristically accepted the *fait accompli* and withdrew the demand for standby wages in view of an understanding that the network would rectify the situation in 1942. Because that understanding was not honored, shortly before the 1942 concert was scheduled to air Petrillo informed NBC that it must cancel the concert or face the loss of scheduled musical programs. In doing so he not only disappointed 160 boys and girls and, no doubt, their families and friends as well; he also left himself vulnerable to new kinds of criticism.[11]

At the behest of Senator Arthur H. Vandenberg of Michigan, the Federal Communications Commission (FCC) asked Petrillo to justify his action. Petrillo replied that he had acted on behalf of the union for the protection of its members. The rule he enforced at Interlochen, he explained, was a standard one that applied to all commercial employment of musicians. "You must remember," he wrote of working musicians, "that this already is an overburdened profession. In some of our locals unemployment reaches 60 percent; in some 75 percent; and in other locals as high as 90 percent." The children at Interlochen were thus unlikely to become professional musicians, Petrillo continued. "After having studied for many years, they will find themselves in a starving profession."[12]

This response was a tactical mistake that Petrillo's critics were quick to exploit. At once editorialists added charges of cruelty to children to those of disloyalty in wartime and dictatorial power over musicians. These charges seemed to gain credence when the children at Interlochen signed a petition protesting what Petrillo had done. Political cartoonists had a field day. Reg Manning of the *Phoenix Arizona Republic* drew Petrillo dressed as Caesar calling children "scabs" and banning them from radio. He also showed "Caesar" bellowing decrees at radio audiences and musicians, and even at Uncle Sam. Interlochen hardly reflected the nature of the problems confronting musicians or the issues involved in the ban on recording. But in the hands of Petrillo's critics and opponents, it became a symbol that trivialized the struggle against canned music.[13]

The Interlochen incident heightened congressional interest in the AFM and the ban on recording. On August 27 Senator D. Worth Clark, a Democrat from Idaho, introduced a resolution calling for an investigation of the union's "acts, practices, methods, and omissions to act," which Clark characterized as threatening "the national welfare, the public morale," and "the public good." The resolution went to the Interstate Commerce Committee, which recommended approval, and when the House approved it, the Commerce Committee named Clark to oversee the investigation. The resulting "Petrillo Probe Subcommittee" promptly summoned representatives of the Justice Department, the FCC, and the Office of War Information (OWI) to testify on the nature and effects of the ban on recording. Their testimony reflected the mounting pressure on Petrillo and the union.[14]

Speaking for the Justice Department, Thurman Arnold told the committee that Petrillo had trespassed legal boundaries. The Interlochen incident, he said, was "a step beyond the closed shop," was in fact an "attack on individual freedom in America." The recording ban, he continued, threatened restaurants, hotels, jukebox operations, and other small businesses simply because they utilized "new inventions for the rendition of music." The issue the ban raised, Arnold told the committee, was whether trade unions could force employers to hire "unneeded labor."[15]

Elmer Davis, head of the OWI and a former news analyst for CBS, seconded Arnold's testimony. Davis insisted that the ban on recording threatened small radio stations across the nation and thereby jeopardized the government's "propaganda broadcast structure." He also suggested that the ban impeded the war effort by lowering military as well as public morale. Field commanders in combat zones, he reported, had complained to him about the shortage of recorded music and the effects of the shortage on their troops.[16]

James L. Fly, chairman of the FCC, was more sympathetic to Petrillo and the union, and to Petrillo's handling of the Interlochen incident. The National Music Camp, he noted, used the children's concert as a form of

Waving his finger like a baton, Petrillo testifies in early 1943 before a Senate subcommittee investigating the AFM's ban on record making. That July, before the War Labor Board, Petrillo questioned federal authority to force musicians back to work and thus into "involuntary servitude." *(AP/Wide World Photos)*

advertising and thus should pay for on-air musicians. In any case, he added, the Interlochen incident was only one of thirty-one instances in which the union had kept amateurs off the airwaves. A recent survey of employment in radio, Fly continued, indicated that recordings did in fact harm musicians, even though the networks themselves relied almost exclusively on live music. The problem was the affiliated stations, which now numbered nearly five hundred and which broadcast as much recorded as live music. Adding to the problem, the three hundred or so unaffiliated stations nationwide relied on recordings for more than 80 percent of the music they played. The FCC chairman pointed out that radio employed only 2,171 full-time musicians, fewer than three for every station in the country. It employed 1,171 part-time musicians, plus 685 "hillbilly" entertainers who also worked on a part-time basis. Fly concluded that broadcasters were paying a small price for musicians, especially in view of their heavy reliance on musical programming and their rising profits. Petrillo could not have said it better.[17]

The Clark committee called no one from the AFM to testify, but it allowed Joseph Padway, the union counsel, to speak on the union's behalf. Padway told the committee that Petrillo was not a dictator but an elected leader "simply carrying out the orders of the AFM." "Wouldn't you say that Mr. Petrillo dominates the union?" Clark asked. "Exactly the contrary," Padway replied. "The AFM is a very democratic organization. I don't think it is less democratic than Congress." Padway scoffed at charges that the recording ban threatened military or civilian morale. The union had agreed to produce recordings for troops abroad, he noted, and was cooperating with the government's Victory Disk Project. In fact, he added, its members recorded "V-disks" free of charge for distribution outside the United States. There was no one and no group more loyal to America or more committed to the war effort, Padway insisted, than Petrillo and the AFM.[18]

It is easy to believe that Clark and the committee had their minds made up before the hearings began. Clark had earlier called Petrillo a "gangster," and the committee report reflected that characterization. The ban and the Interlochen incident were serious attacks on public well-being during wartime, Clark wrote, recommending a congressional investigation of the AFM. The Senate accepted the recommendation and set aside $5,000 for a larger probe to lay the basis for whatever legislation might be appropriate to curb the union and its president.

This action signaled a changing mood in Congress. By the early 1940s the attitudes of federal officials toward organized labor had begun to shift from the liberal, pro-union stance of the early New Deal to the conservative, antiunion posture that emerged after the war. Many in Congress had never supported the labor reforms of the New Deal and now saw opportunities to reverse them. Public reaction against the recording ban and the Interlochen incident provided the occasion for this shift rather than its cause. The standby rules enforced by the AFM, like similar rules in the railroad and trucking industries, had already drawn the attention of an increasingly conservative Congress. At a time when the government had drafted millions of men and labor shortages threatened war production schedules, many lawmakers considered "featherbedding" an offense against the general welfare. In this atmosphere the ban on recording became the focal point of a public debate on the whole "make-work" issue, and *Broad-casting* predicted that Congress would pass legislation to "fit the circumstances" of this indefensible situation. Labor leaders, the journal noted, were "plainly worried."[19]

On October 12, 1942, the trial of the antitrust suit brought by the Justice Department against the AFM opened in Chicago. There, before Judge John P. Barnes, Thurman Arnold challenged the legality of union contracts that required employers "to maintain obsolete or inefficient methods." Could unions, Arnold asked the court, force employers to "refuse to introduce new mechanical improvements in order to compel the hiring of unnecessary labor"? Arnold thought not, obviously implying a parallel between Petrillo and Ned Ludd, an earlier enemy of "progress."[20]

In taking this position Arnold was measuring progress in terms of economic efficiency, while Petrillo and the union were measuring it in terms of job opportunities. Arnold told the court that there was little unemployment among musicians. His own study of the radio industry, he said, indicated that most men counted as unemployed musicians in small cities with affiliated and unaffiliated radio stations had moved elsewhere or taken jobs outside of music, and were thus unavailable for musical employment. Those who remained in those cities, he added, were mostly amateurs unqualified for radio work.[21]

A variety of witnesses buttressed Arnold's testimony. Neville Miller, president of the NAB, testified that the AFM was "engaged in a campaign to compel the paying of tribute by broadcasting stations . . . whether the stations can employ or utilize the services of union musicians or not." More than half the stations employing musicians on a weekly basis, Miller said, were paying more of them than they needed. He also urged that the recording ban be lifted for the sake of consumers. "Before the phonograph record," Miller explained, "only persons who could pay to go to the large concert halls in the large cities could hear the great symphony orchestras, and only the persons who could afford to go to fashionable restaurants and hotels could hear the best dance orchestras." If the recording ban succeeded, those days would return. The musicians' union, he added, "cannot expect the American public to stop listening to the artists whom they have learned to enjoy and [instead] listen to the small aggregations of part-time non-professional musicians who are available for employment in the small communities."[22]

Edward C. Coontz, who owned station KVOO in Tulsa, Oklahoma,

Stage to Studio

personalized these ideas. In an effort to highlight local programming, Coontz had recently tried to hire local musicians but had been unable to find talented individuals or groups to fill the schedule he had planned. "We have tried combination after combination of musicians," he told the court, "but have, as yet, not found one that is satisfactory enough for general use. Last Spring," he continued, "we attempted to get an orchestra from the union for one of our best local accounts." He auditioned musicians "even down to the point where their four top men were seated and asked to play a simple hymn written in four parts. They couldn't get together on it, so the idea was abandoned." Harry Le Poidevin, owner of station WTAR in Racine, Wisconsin, offered another perspective, that of the owners of even smaller stations in even smaller cities. The musicians' union in Racine, Le Poidevin testified, "cannot offer the same high grade of music" provided by recordings and network programs. "Local advertisers do not desire to sponsor musical programs unless they are of exceptional quality," he continued, and "local bands and local individual musicians, with rare exceptions, cannot compete with the top-notch musicians of the country."[23]

Jukebox operators offered similar testimony. Al Dolins, to illustrate, told the court how the recording ban threatened his business. Dolins had machines in diners, restaurants, taverns, and army camps throughout Massachusetts and Rhode Island, which he serviced once a week, collecting coins and updating record selections. Few of the locations he serviced, he testified, had ever employed musicians. "In my opinion," he said, "these locations do not warrant musicians and would probably never have them if the automatic phonographs were removed." But "if the current supply of records were stopped," he added emphatically, "I would not be able to continue my business." Dolins estimated his investment in his business at $100,000 and noted that he and his ten employees would be out of work if his business failed.[24]

A. L. Pressley of Pickens, Mississippi, gave the court the perspective of proprietors who rented jukeboxes from entrepreneurs like Dolins. Pressley owned and operated the Grapes Camp Tavern on Highway 51 north of Pickens. The tavern consisted of a lunchroom with counter service and an adjoining dining room with tables and booths and a dance floor. Each room had its own jukebox. Because of the small scale of his operation, Pressley told the court, he could not afford to hire live musicians, and his business would be seriously harmed if the recording ban continued. "I have in the past tried to throw special dancing parties and used an orchestra therefor," he said. "However, I found this to be impractical and soon had to discontinue the practice. Even though I made a small cover charge, the expenses far outweighed my receipts." Pressley was therefore "dependent on the phonograph to serve as an inducement to bring those people to my place who are out for a good time and some recreation."[25]

Testimony in the antitrust suit showed clearly the implications of the recording ban for business and consumers. It showed, for example, that big and small business alike opposed the boycott. That in itself was important. The recent expansion of big business into different sectors of the leisure industry threatened small firms everywhere and had sometimes produced open conflict. Large and small firms, however, put aside their differences to oppose the recording ban because both stood to lose if the ban succeeded. "Without these mechanical records," a radio station owner explained, "the quality of our programs [would] deteriorate" and music consumers would suffer.[26] As the president of the NAB told the court self-servingly, music lovers had become accustomed to hearing the best musical performances in the comfort of their own homes for what amounted to nominal costs. The recording ban threatened that custom, too.

Not all of the testimony was adverse to the interests and concerns of the AFM. Some of it also showed clearly that new recording, transmitting, and amplifying technologies had not benefited musicians as a group. Arnold was right on this point: thousands of musicians had moved from small towns to media centers, and most of those who remained found it necessary to find other jobs. That was among the most important consequences of the sound revolution that had transformed the business—and the work—of musical entertainment. The centralization of job opportunities had forced thousands of musicians to relocate in order to live by their musical skills, and those who remained at home found their skills devalued as well as their prospects for work diminished. The testimony of the broadcasters from Tulsa and Racine repeated what was common knowledge: the high quality of performances on network radio and first-class records made locally produced music sound unsophisticated, if not amateurish. The preference for recorded music and network programs was therefore understandable, whatever its implications for working musicians outside media centers. Big-city bands featured exceptionally talented musicians, and their members played in acoustically superior settings. The most popular bandleaders and singers typically fronted these bands, which had the time and money they needed to rehearse for as long as necessary. Few local musicians could compete with such groups.

The AFM's chief counsel, Joseph Padway, expected to testify in the Chicago case. Toward that end he secured a statement from the AFL denouncing the stance of Arnold and the Justice Department as "pernicious" and "heartless" toward American workers. But Padway had no chance to testify. At the conclusion of the government's case Judge Barnes dismissed the suit on the grounds that his court lacked jurisdiction over it. This was a labor dispute, he explained, to be resolved according to provisions of the Wagner Act. In making his ruling Barnes also made his own sentiments clear. "This is a controversy between masters and servants," he said, "a question [of] whether the servants must make music as the masters direct." In his closing comments Barnes disputed Arnold's argument that no unemployment existed among musicians. He estimated that half of the former musicians who now had other jobs had those jobs because they "were not able to make a reasonable living in their chosen field."[27]

Barnes's dismissal of the suit caught AFM leaders by surprise, perhaps because federal judges generally supported employers over trade unions. Petrillo was jubilant, and understandably so. Barnes had unequivocally sustained the union's right to continue the recording ban. "The ban still stands," Petrillo told the press. Barnes's ruling showed "that the abuse of a high labor official cannot triumph over justice and labor." Padway seconded Petrillo's view. The court made it "crystal clear," he said, that musicians were "in the right" in the recording-ban dispute, while Arnold, "the champion of big business," had learned that he could not force the working musician "to erect the gallows on which he is to be hanged." Arnold, however, appealed the ruling.[28]

The verbal sparring that characterized the public discourse between Petrillo and his critics showed the degree to which labor conflict no less than other forms of social disputation is bound up in language. Through the conscious and unconscious manipulation of verbal symbols, musicians and their employers alike struggled to affix social meaning to the sound revolution and to their differing responses to that revolution. They fought, that is, for control of the meaning of the communicative symbols through which the public perceived their struggle. In the language of the poststructuralists, they vied for the power to assign signifieds and even referents to the signifiers they bandied about. In doing so neither side willfully distorted the past or the present, but both explained their struggle in the same way that today's readers will "read" it: according to their respective circumstances, interests, values, and understandings of what constituted social reality and the social good. For these reasons employers found Thurman

Arnold's antimonopoly rhetoric especially resonant, while the union found Judge Barnes's words equally appealing.[29]

THE CHICAGO VICTORY came while Petrillo was preoccupied by a dispute with radio stations affiliated with the networks. Although some 250 of those stations employed staff orchestras, many did so grudgingly and challenged local unions at every turn over the continuing requirement that they do so. Between late 1942 and the summer of 1943 Petrillo intervened in disputes across the country that resulted from this situation. Whenever union locals appealed to him in such disputes, Petrillo responded by threatening to pull instrumentalists from network programs.

He made good on his threats whenever he felt it necessary to do so. In January 1943 he banned remote broadcasts on CBS and NBC's Blue Network because of a dispute in Pittsburgh. Local 60 and H. J. Brenner's stations there had again locked horns, this time over Brenner's effort to shorten the staff orchestra's employment season. Angry at the resulting ban on remote broadcasting, Mark Woods, vice president of NBC, complained that the action not only "penalized" the network for "conditions beyond its control" but also punished all network affiliates simply "because one Blue affiliate has differences with the musicians' organization." But it was that effect that made Petrillo's tactic work; and in accord with that tactic, Petrillo rescinded the Pittsburgh ban on January 26, when orchestras at Brenner's stations "were employed in conformity with previous contracts." This and similar actions elsewhere showed that Petrillo was unwilling to give up any union foothold in radio. Indeed, the repeated confrontations hardened his resolve to maintain the recording ban, for it was broadcasters like Brenner who wanted to substitute recordings for the services of local musicians.[30]

While Petrillo worked to save jobs in radio, lawmakers investigated him and his tactics for possible violations of antitrust and other laws. Accepting the previously noted recommendation of Senator Worth Clark of Idaho, the Senate formed a special committee to conduct the investigation. The committee made the well-known Washington attorney Herbert M. Bingham its chief counsel, and one of Bingham's first acts was to summon Petrillo, who thereby gained the dubious distinction of being the first American to have to defend his actions as a labor leader before Congress.[31]

For two days Bingham and the senators grilled Petrillo about the recording ban. What precisely did he and the union want and expect the ban to accomplish? A candid answer—more jobs for musicians in radio and elsewhere—was impossible, for that would make the ban a secondary boycott. In a series of evasive but revealing responses to this and related questions, Petrillo stated in effect that what musicians wanted was more work. Musicians would go back to recording immediately, he said, if the industry gave them a "fair share" of the profits of their work. The solution to the problem represented by the ban must therefore come from broadcasters and recorders, not musicians. Yet Petrillo had solutions of his own. The ban could be rescinded, he implied, if recorders withheld their recordings from stations employing fewer than the requisite numbers of musicians (which would seem to be clear evidence that the ban was in fact a secondary boycott). Or failing that, Congress might enact legislation giving the union property rights in recordings. Royalties from the sale of records could then be used to create jobs for musicians.[32]

In calling Petrillo to testify, lawmakers gave the fiery labor leader a public forum, which he used with consummate skill. Like his counterpart at the NAB, Petrillo played on the sentimentalities and the anxieties of Americans as voters and musical consumers. He especially exploited public concerns about monopoly and the inordinate power of big business. He argued that the campaign against him and his union evidenced the dangers as well as the consequences of concentrated ownership in industry. What should be investigated, he said, was not him and his union but the music industry, over which "a few giant corporations" exercised "tremendous control" to maximize their profits "at the expense of the live musicians." Petrillo could also wrap himself and his union in the American flag. He reminded senators that the AFM and its members had bought thousands of dollars worth of war bonds and that twenty-five thousand of its members were in the armed forces. Refuting charges that the recording ban was unpatriotic, Petrillo noted that musicians had made hundreds of victory disks for distribution to the armed forces and had done so without charge, and that he and they had agreed to end the recording ban if President Roosevelt found that it was undercutting troop morale. In short, Petrillo seized an opportunity to reinterpret the recording ban in a way quite different from the one then circulating in the press, and different as well from the one radio had previously presented to the public.[33]

Petrillo's testimony was effective enough to confound his critics. The union leader responded to questions with evident candor and self-effacement as well as wit; perhaps he even changed the minds of some senators. In any case the committee, upon the advice of chief counsel Bingham, recommended no legislation against the union or the recording ban. Even Petrillo's adversaries acknowledged the effectiveness of his performance before the committee. *Broadcasting* conceded that Petrillo "made a far better witness than was anticipated. We understand," one of its editorialists wrote, "why the AFM elected him president."[34] But Petrillo had also made a significant concession at the hearings. Probably out of perceived necessity, he had agreed to come up with a plan to end the ban. To skirt the problem of the secondary boycott, his plan could make no specific demand on radio. He evidently decided therefore to sacrifice one goal (more jobs in radio) for another (financial concessions from the recording industry). In doing so he and his advisers may have concluded that employers could survive the ban much longer than the union had assumed. After all, the ban was six months old, and the industry showed no signs of conceding anything.

YET THE INDUSTRY did show signs of discord. As record supplies thinned, small businesses without the resources and diversity of industry leaders had begun to talk of compromise. The tension generated by this emerging division was evident as early as October 1942, when Samuel R. Rosenbaum, head of station WFIL in Philadelphia and former head of the Independent Radio Network Affiliates (IRNA), criticized the industry-backed press campaign against Petrillo as a "masterpiece of ineptitude." Rosenbaum believed that broadcasters were poorly served by "labor-baiting and labor-leader smearing" pronouncements that were "a relic of a past generation. With the entire press of the United States at our disposal, and with powerful branches of Government lending themselves amiably to the effort," Rosenbaum told industry leaders, "all we have been able to think of is to attack the integrity and personal characteristics of one labor leader."[35]

Rosenbaum was not the only one expressing such sentiments. Differences within the industry over responses to the recording ban were the subject of a *New York Times* story in late 1942. According to the story, some network affiliates were "out of sympathy" with the stance taken by the NAB toward the ban. Management at those stations, the *Times* reported, thought the solution to the ban would come not through legal challenges or public relations campaigns but through direct negotiations with Petrillo. At least one broadcaster found "merit" in the union argument that "stations using music all day ought to pay something to musicians." Such opinions undoubtedly worried industry leaders, who like musicians recognized the importance of solidarity. They also recognized that manufacturers who came to terms with the AFM concerning the ban would reap immediate competitive advantages. These signs of disunity worried industry leaders no less than they encouraged Petrillo and the musicians.

The NAB held fast. Sidney Kaye, counsel for the association during the crisis, denounced Rosenbaum as someone who "does his thinking in an ivory tower." Asked about the possibility of settling the ban, Kaye responded, "We don't want to settle it, when we didn't start it. Instead of giving up, we are going to fight it out." To control the damage caused by Rosenbaum's comments, executives of NBC, CBS, and Mutual wrote NAB president Neville Miller affirming their support of the stand against the union. "We feel," wrote Paul M. Kesten of CBS, "that the activities of NAB are proper . . . and we have no desire to do anything other than to support your position." Frank E. Mullen, Kesten's counterpart at NBC, added, "We have confidence that your association is handling the matter in the interests of the industry and of the public."[36]

The rift among employers widened after the Senate hearings ended. The ubiquitous Rosenbaum, now a symbol of compromise, continued to criticize the NAB, telling the press that musicians "are entitled to fair protection against free exploitation by commercial users of records made for home use." Rosenbaum endorsed the union proposal that broadcasters, recorders, and jukebox operators pay a fixed fee to the union as compensation for recording musicians. He also proposed that the jukebox industry pay 4 percent of its net profits to the AFM, thus giving the union $6 million a year "for the employment and encouragement of live musicians." Rosenbaum thought that under this plan affiliated stations, which then spent up to 5.5 percent of their profits on musicians, might lower that expenditure to 2 percent, or even 1 percent.[37]

Petrillo seized the opportunity Rosenbaum's proposal opened. Acknowledging his pledge to present a proposal to end the ban, Petrillo wrote the recording companies embracing the fixed-fee principle. Specifically, he proposed that recording companies pay the union a "royalty," the amount of which was negotiable, that the union would use to create jobs for musicians. The proposal was a turning point in the dispute. The union had offered a specific plan for ending the ban, one that appeared to be legal and that some employers accepted. But it was not unlike the proposal motion-picture producers had not so long ago rejected.

Representatives from RCA-Victor, NBC, CBS, Muzak, Decca, and half a dozen other companies met in New York in late February to discuss Petrillo's proposal. At the conclusion of the meeting the NAB issued a statement denouncing the proposal as one that embraced a "startling new kind of social philosophy," one that industry leaders found "dangerous and destructive." The plan was also "socialistic," an effort to generate a "private relief fund" for a group of workers most of whom were no longer unemployed. Industry leaders thus flatly rejected Petrillo's proposal. The union responded to the rejection—and to the criticism of the proposal—in kind. On March 17 the executive board wrote industry leaders that their summary rejection of Petrillo's proposal violated the letter as well as the spirit of the collective bargaining process by "fail[ing] to consider proposals in good faith." The board then vigorously defended the union's social philosophy. "Those who benefit from the displacement of human labor," its letter read, "should share the burden of the cost to the displaced workers."[38]

The ideological positions of the two sides seemed irreconcilable. On the one hand, employers welcomed the social benefits as well as the profits from advances in musical technology while denying that the use of recordings in broadcasting was detrimental to musicians. On the other hand, the union insisted that the radio and recording industries were reaping windfall profits by utilizing musical labor and technology in ways that displaced musicians. The union insisted that employers share the social cost of this result, while employers refused to acknowledge that the consequences of technological change had social implications. These opposing assessments underscored the fact that labor and management had different measures of progress. Management saw the fixed-fee proposal in terms of economic cost: if implemented, it would increase production costs and reduce profits. Musicians, on the other hand, saw it as one way of dealing equitably with technological changes that were already having devastating consequences for a large segment of the working class.

The differing outlooks did not preclude continued bargaining. The NAB proposed a new round of negotiations, which began on April 15, 1943. The mood at the meetings was surprisingly cordial; but of course mood alone does not settle labor disputes. Manufacturers of popular recordings and long-playing transcriptions still rejected the fixed-fee plan, so Petrillo made a counterproposal of something he had always wanted. The union, he proposed, would end the recording ban if the recording companies would agree to withhold their records from radio stations the union deemed unfair to labor. This would give the union the power to force affiliates to employ minimum-size orchestras by threatening to deprive them of new recordings. The NAB promptly labeled this as a call not only for an illegal secondary boycott but for "business suicide" as well, since it would give the union a "stranglehold over independent stations."[39]

The adamancy of this response belied the dynamics of a now rapidly changing situation. Industry leaders could no longer count on small companies to follow their lead, as the reaction to Rosenbaum's proposal made clear. The government too appeared unable or unwilling to curb Petrillo, especially after the Supreme Court affirmed the union's right to continue the recording ban by sustaining Judge Barnes's ruling in the Chicago antitrust suit.[40]

Still, industry leaders refused to compromise. In a last-ditch effort to end the ban without concessions, the NAB took its case to the War Labor Board (WLB), which could intervene in any industrial conflict that jeopardized the war effort. The move was an act of desperation. The WLB was a product of the three-way agreement reached by government, management, and labor early in the war to ban strikes and lockouts that adversely affected the war effort. Its decisions were advisory only and their effectiveness entirely dependent on voluntary cooperation.[41]

As the WLB studied the NAB appeal, events made the appeal itself moot. On September 30, 1943, the Decca Recording Company and its subsidiary transcription division, World Broadcasting, signed a four-year agreement with the AFM accepting the fixed-fee principle. Because Decca produced nearly a quarter of all records sold in the nation, its action forced other recording companies to make similar agreements. Within three weeks four large transcription companies had done so—Langworth Feature Programs, Standard Radio, C. P. MacGregor, and Associated Music Publishers—and by January 1944 some fifty additional companies had followed suit. That number doubled in the next few months.

All of the companies, including Decca, agreed to pay the newly created AFM Record and Transcription Fund a royalty of between a quarter of a cent and 2 cents for each record they sold, depending on the size and price of the record. The AFM would in turn distribute the fund to its locals according to a formula based on the size of membership. The locals would use the money to finance concerts that were free to the public but for which the musicians received pay at union scale. At last, musicians would have a source of income to replace, at least partially, what they had lost from the advent of talking movies, radio broadcasts of recorded music, the demise of vaudeville, and the unexpected popularity of jukebox music. From Tacoma to Tallahassee, union locals could look forward to several thousand dollars a year from the Record and Transcription Fund.[42]

The settlement by Decca and other small companies was probably inevitable once it became clear that union solidarity was firm. In labor disputes as in other things, employers cooperate as long as cooperation serves their individual purposes. When some of them see advantage in coming to terms with a union, their common front often breaks down. The recollections of Milton Gabler, who worked in Decca's Artists and Repertoire Department from 1941 to 1971, testify to this general pattern. Gabler recalled that Decca president Jack Kapp "wanted to do business and make money, [and in 1943] saw a chance to get the jump on RCA and Columbia." He also recalled that the two larger corporations, unlike Decca, were concerned about the implications of the fixed-fee plan for labor costs in radio. "RCA and Columbia were afraid if they gave in to Petrillo too much they would have trouble with the next contract negotiations for their orchestras."[43]

In brief, Decca and other companies that signed the fixed-fee agreements came to see that their interests were not the same as those of such industry giants as RCA-Victor and the recording divisions of NBC and CBS. Small companies lacked the diversified resources the much larger ones had, and they realized that signing the fixed-fee contracts would significantly improve their competitive positions. In fact, when Petrillo lifted the ban, Decca became the nation's largest record manufacturer. Popular singers and bandleaders under contract with other companies soon switched to Decca, which further strengthened its competitive position. Decca's signing of the renowned violinist Jascha Heifetz, for example, produced a windfall of publicity and profits. One of the reigning masters of the violin, Heifetz had previously worked for RCA-Victor.

This course of events showed again that American business is not a monolithic entity. There were in fact differences of interests and perspectives capable of splitting businesses in the music industry into competing groups under given circumstances. Yet the division that emerged represented pragmatic, ad hoc interests rather than differences over the desirability of strong and effective labor unions. The owners of small and large businesses in the music industry remained committed to the values and objectives that had united them in resisting the recording ban at the outset. If they differed two years later over how and when to settle with the union, the difference was due to the differential impact of a continuing ban on their individual positions within the industry. There was no difference among them over such fundamental things as the proper organization of society or the cause and consequence of the inequalities inherent in capitalist economies.

The partial resumption of record production was a major victory for the AFM. The action of Decca and the other early signers of the agreement with the union increased the pressure on industry leaders, who watched their competitors capitalize on what amounted to a new system of industrial relations in music. The new agreements notably benefited the union and strengthened the position of its leaders not only within the union but vis-à-vis their counterparts in industry. Musicians who lost income during the recording ban returned to work with much clearer perceptions of the value of collective action and solidarity in the face of adversity. The new contracts satisfied union locals that had long demanded aggressive action in behalf of musicians. The Record and Transcription Fund rejuvenated hundreds of locals whose ability to act in behalf of their members had eroded markedly during the preceding decade.

Yet the union's victory was far from complete. The new fixed-fee contracts said nothing about the use of recordings by commercial enterprises deemed unfair to labor. Nor did industry leaders relax their opposition to fixed fees. On the contrary, all of them continued to use stockpiled records, clinging to the hope that the federal government would rescue them from union diktat. Specifically, they hoped the War Labor Board would contest the fixed-fee contracts on the grounds that they represented wage adjustments, and thus violated wartime wage stabilization criteria. Some even hoped the board would ask President Roosevelt to use his emergency powers to rescind the recording ban in order to boost wartime morale.

The WLB considered the NAB challenge to the ban in early 1944. In March the hopes of broadcasters rose when a WLB panel recommended that the board overturn both the recording ban and the fixed-fee contracts. But on June 15 the board itself rejected that recommendation. The Decca contract, the board ruled, did not require government approval "since the payments to be made . . . are not wage adjustments within the meaning of the wage stabilization program." The board then ordered the union to lift the recording ban, while directing the recording companies to "compromise" with the union and "reach an agreement regarding the amounts and the schedule of escrow payments to be made." Companies that failed to follow this directive would have the amounts and schedules set by the board.[44]

The agreements thus mandated did not materialize. Petrillo had always insisted that the WLB had no jurisdiction in the dispute, and in negotiating with industry leaders he steadfastly rejected terms that differed from those accepted by Decca and the other early signers of the accord. He argued in fact that the contracts with those companies obligated him to insist on the terms of those contracts. Petrillo also ignored the order to lift the recording ban. To do that, he said, would allow the unsigned companies to stockpile recordings and hold out indefinitely against the union. The companies were equally unyielding, and the ban against them dragged on through the summer of 1944.[45]

The hopes of industry leaders probably rose in the fall of that year, when the WLB and the Economic Stabilization Board sent the "canned music controversy" to the White House for resolution. In early October, Roosevelt asked Petrillo to lift the recording ban "in the interest of orderly government." Petrillo's failure to comply with the WLB directive against the ban, Roosevelt argued, would "encourage other instances of non-compliance" and reverberate to the detriment of the national interest. "What you regard as your loss," Roosevelt intoned, "will certainly be your country's gain." Petrillo demurred. In a tactfully worded response, he told the president that he was obligated to honor the fixed-fee contracts he had signed, which contracts made it "illegal" for him to give "recalcitrant companies different [and more favorable] terms." Roosevelt thereupon promised to "look into the law" concerning the ban, but he did not order Petrillo and the musicians to end it. This additional victory for the union left the holdouts no choice but to come to terms with Petrillo.[46]

Other factors encouraged them to make that choice. Since 1942 record manufacturers had had difficulty procuring shellac, a heat-resistant material, imported from India, that was necessary for the manufacture of phonograph records. Shellac protected the surface of finished records and muted surface noise when they were played; it also constituted about 20 percent of the material in the records themselves. Wartime restrictions limited the availability of shellac, raised the cost of record manufacturing, and thus discouraged record production. In 1944, however, the War Production Board began lifting restrictions on the importation of shellac, thereby permitting manufacturers to increase the volume of record production without raising retail prices. This meant, among other things, that Decca and other companies with fixed-fee contracts could increase their share of the record market.[47]

Perhaps this was the straw that broke the camel's back. On November 9, 1944, the industry leaders capitulated. "I received a telephone call from an official of one of the companies," Petrillo later recalled, "asking me to come to New York in order that they might sign contracts with the Federation." Two days later, representatives of RCA-Victor, NBC, and CBS

President Franklin D. Roosevelt's failure to order an end to the recording ban (he asked Petrillo to send musicians back to work; Petrillo kindly refused) produced this biting Rube Goldberg cartoon for the *New York Sun* of October 6, 1944.

signed four-year contracts obligating them to make contributions to the Record and Transcription Fund on the same terms as Decca and other companies that had already signed contracts with the union. Like those early signers, the corporate giants pledged to pay into the fund specific sums for each record they produced, depending on the size and price of the record.[48]

Indulging his customary penchant for hyperbole, Petrillo called the settlement "the greatest victory for a labor organization in the history of the labor movement." There was at least some basis for the claim. Not just the musicians' union but trade unionism had taken a step forward with the settlement. No union had ever before forced employers to contribute to a fund designed to provide jobs and income for workers displaced by technology. In signing the fixed-fee contracts, recording companies had acquiesced in, even if they did not positively agree with, the principle that technological change imposed social costs that employers had a responsibility to share. The social implications of the contracts became clearer after the war, when workers in the automotive, railway, coal-mining, printing, and other industries negotiated similar arrangements to cushion the impact on workers displaced by technological change. Employer-supervised retirement plans, unemployment benefits, and workmen's compensation plans are examples of the "welfare privatization" that surfaced in these industrial arrangements and reflected the social principle embodied in the Record and Transcription Fund. In each of these schemes employers who benefited from technological advances in industrial production obliged themselves to help displaced employees.[49]

Employers did not willingly accept this obligation. Those in the recording industry in fact mounted an expensive public relations campaign to discredit the effort to make them do so. That campaign embittered Petrillo and left him defiant. "Instead of showing friendliness," he said of industry leaders on the day he signed the agreements with them, "they have displayed bitterness, unfairness, injustice, trickery and reactionism which would do justice to the slave owners of pre–Civil War days. . . . They substituted for the ordinary, usual and fair processes of collective bargaining a campaign of mud-slinging, dirt-throwing and false propaganda." Their actions in the campaign had been "vile, indecent, malicious, and filthy," and Petrillo warned that if they violated the obligations they now agreed to, he and his union would "not hesitate to break off relations and leave them to die by their own nefarious schemes."[50]

This rhetoric obscured the fact that the outcome of the recording ban was less than the complete success Petrillo claimed it to be. Petrillo had hoped to restrict the commercial use of recordings and in doing so create thousands of new jobs for musicians. He had especially wanted to create jobs in radio, or at least secure the jobs musicians already had in that industry. To that end he had tried to force recorders to withhold records from stations without staff orchestras. But he had to abandon that effort because it constituted what the law called a secondary boycott—an illegality. In addition, by 1943 Petrillo's own advisers were conceding that recordings of national artists were superior to live performances of local talent and were thus rightly preferred by radio audiences. Petrillo himself acknowledged that the state of recording technology was such that it made "a second class band sound like a first class one." The substance behind that admission might be the real reason Petrillo and his advisers decided to settle for partial victory in the war against recordings.[51]

Seven

Balancing Success and Failure

DURING WORLD WAR II the American Federation of Musicians (AFM) won important concessions from employers, but after the war the labor history of musicians, like that of other workers at the time, was stormy. Industrial conflict intensified as the war ended. In the two months following V-J Day, the number of workdays lost nationwide to work stoppages skyrocketed. Across the country hundreds of thousands of miners, machinists, longshoremen, steelworkers, truck drivers, and other workers walked off their jobs, sometimes in defiance of their own unions. The resulting conflict peaked in 1946, when approximately 4.6 million workers found themselves involved in strikes.

To unionized workers, at least, this unprecedented activism was justified, even overdue. They—and nonunionists too, though it is difficult to generalize about them because they have been so little studied—had made major sacrifices during the war, enduring extended hours, hazardous conditions, and uncompensated speed-ups while honoring no-strike pledges and wage freezes. Yet neither the public, the government, nor employers showed much appreciation for the sacrifices workers had made. When the government lifted price controls at the end of the war, wage controls remained in force, which allowed living costs to rise while income remained flat. When wage controls did end, employers resisted wage increases even though many of them had reaped huge profits from war-related contracts. At the same time, industrial and other war-related employment plummeted as the government canceled those contracts before industries had retooled for peacetime production and millions of discharged soldiers were returning to the civilian workforce. The net effect of these developments was to decrease job security and lower real income for vast numbers of workers.

Responding to one of the largest and costliest strike waves in the nation's history, lawmakers began to reexamine New Deal labor legislation, which many of them thought was at least partly responsible for the unprecedented levels of labor unrest. Across the country newspaper headlines condemned the strikes and the unions that had called them, and public opinion became increasingly antiunion if not antilabor. By the time of the 1946 congressional elections, public as well as congressional opinion had shifted decisively against unions and the New Deal coalition that had supported them. President Roosevelt, who had generally sympathized with organized workers, was succeeded in the spring of 1945 by Vice President Harry Truman, whose attitudes toward labor were unknown and thus problematic, and whose political clout was limited. In the postwar climate of opinion, however, even Roosevelt would have been hard-pressed to defend the interests of labor because the depression-era coalitions that had passed the Wagner Act, the Fair Labor Standards Act, and other pieces of pro-union and pro-labor legislation had been steadily eroded in the congressional elections of 1940, 1942, and 1944.

This changing political climate had major implications for musicians and their union. Indeed, it presented the most serious challenge they had faced since the advent of sound movies. The growing influence of the federal government in industrial relations meant that the future of the Record and Transcription Fund as well as of musicians in radio depended upon the government. The fact that the changing political mood coincided with advances in broadcasting technology increased the significance of that dependence. The spread of FM radio broadcasting and the beginnings of television raised hopes for new employment, but those hopes depended on the willingness of Congress to enact laws enabling musicians to protect musical employment in those industries. Against the backdrop of evolving government-business relations in the post–World War II years, musicians continued their efforts to safeguard what they already had while trying to exploit further changes in broadcasting technology.

WHEN WORLD WAR II ended, American musicians had been coping with the effects of rapid technological change for two decades and more.

Renowned composer Dimitri Tiomkin, his back to the camera, conducts studio musicians playing the score for *Duel in the Sun,* a David O. Selznick production of 1946. After World War II some 250 musicians held full-time jobs in motion-picture studios; another 2,500 worked in radio. *(Dimitri Tiomkin Collection, Cinema-Television Library, University of Southern California)*

Despite the loss of theater employment, they could point to meaningful accomplishments in other areas. Their wages in the record industry now exceeded $2 million a year, and industry payments into the record-royalty fund generated another $1.5 million. The motion-picture industry now provided full-time employment for 250 instrumentalists and part-time work for another 5,000. The aggregate annual income for these musicians was perhaps $2.5 million. Income from radio was even more impressive. There, musicians had twenty-five hundred full-time jobs and a number of part-time jobs that cannot be determined precisely; instrumentalists in radio earned more than $21 million a year.[1]

Ongoing changes in broadcasting fueled hopes for even more employment in radio. During the 1930s the advent and subsequent spread of static-free, frequency-modulation broadcasting and the first experimental television transmissions not only fascinated the public but raised the hopes of musicians too. The commercial potential of both of these media was clear by 1940, but World War II delayed its realization for a decade. When the peacetime economy returned, however, entrepreneurs began to capitalize on these technologies. By the end of 1944 a number of FM radio stations and the first few television stations had appeared in media centers, and musicians watched both developments expectantly. Would FM and television make conventional radio obsolete, as some musicians feared? Or, as others anticipated, would they encourage the revival of vaudeville and with it new musical employment? Did FM jeopardize radio orchestras? Or did it promise to compete with AM radio and thus open more jobs for musicians? And what strategies should the union pursue in these rapidly changing times?

Entrepreneurs were better prepared than musicians for the new technological developments. RCA and CBS invested heavily in FM and television technology from the outset and clearly intended to use recorded music in both. NBC, a subsidiary of RCA, had expanded its library of recorded music in anticipation of the programming needs of the new media. The library included more than ten thousand recordings, neatly catalogued according to programming usage. Those with the general designation of "Dramatic Atmosphere," for example, were subdivided into categories labeled "Aftermath," "Haunted House," "Snow Scene," "Motif for Murder," "Stop Press," and the like; those earmarked "Fanfare" were similarly broken down into "Big Moment," "Exhilaration," "Majestic," "Light Atmosphere," "Shopping Center," and other categories.[2] The fact that the networks did not consult the AFM concerning staff orchestras for the new media was a portent of their intentions.

AFM leaders foresaw a bitter struggle over the employment of musicians in FM radio. In 1944 the new chairman of the Federal Communications Commission (FCC), Paul Porter, who had been a lawyer for CBS, reversed the commission policy that forbade simultaneous broadcasting over AM and FM channels. Musicians opposed such duplication because it discouraged the rise of an independent FM radio, which they hoped would compete with AM broadcasts for sponsors and audiences and thereby generate new employment for musicians. Porter's ruling promised to make FM an extension of AM rather than an alternative to it, in which case the networks, given their resources, would dominate FM broadcasting. Porter's ruling allowed, among other things, network sponsors to send the same advertisement over the two media simultaneously at little or no extra cost.[3]

Since management and government promised little help on these matters, the union would have to help itself. In September 1944 the AFM told broadcasters they must obtain a license from the union to transmit the live music of union musicians over FM channels. The purpose of the directive was to halt duplicate broadcasts of AM programs on FM stations. Network executives responded with a flurry of protests. "There is no extra effort required of musicians for FM broadcasting," NBC president Niles Trammell told Petrillo; and in any case AM programs were sent to FM affiliates "without any additional charge to the advertisers" or profits to broadcasters. CBS vice president Paul Kesten asked for an immediate meeting with Petrillo concerning the union directive, so that there would be "no misunderstanding regarding this situation."[4]

Petrillo replied curtly. "Your understanding of the entire FM matter is erroneous," he told Trammell. "There is no misunderstanding on the FM situation," he likewise told Kesten. "The plain facts are that no one received permission to use members of the AFM for FM broadcasting." Petrillo refused even to discuss the issue until the broadcasters agreed to allow the union to license FM broadcasts of musicians. "I will only meet with the networks," he said, "when I am advised . . . that members of the AFM are not being used for FM broadcasting purposes."[5]

While Petrillo was thus working to safeguard the interests of musicians in FM broadcasting, he was endeavoring to do the same thing in television. For two years during the war, when the new and untested medium was in its infancy, the union had permitted musicians to perform in television experiments at wages of $18 an hour. Only a few had done so. But as the radio networks tightened their grip on the fledgling television industry, union leaders began pressing for assurances that musicians would benefit from the medium. Accordingly, in February 1945 the union announced that its members would "not play for Television in any form until further notice." Petrillo then told the networks that he wanted a clear idea of the impact of television before he committed musicians to specific wages and working conditions. "Television," he said, "is not going to grow at the expense of the musicians."[6]

The rank and file let Petrillo take the initiative in these matters. "They backed it up," one union official said of Petrillo's action concerning television; "there wasn't any resentment from the membership." But such statements can mislead. In 1945, when the future of FM and television was unclear, most musicians thought little of Petrillo's initiatives. "[They] didn't mean much to me," one working musician recalled. Even those most likely to benefit from the new media expressed ambivalence about the prohibitions. Studio guitarist Roc Hilman, who worked in television in the 1950s, explained that in the 1940s "television was so new that it wasn't too important." Trumpeter Bob Fleming, who traveled with Kay Kyser's orchestra in 1945, agreed. In that year, he said, "as long as I played well, that's all I was interested in." But Petrillo and other union leaders had clearer perceptions of the commercial implications of the new medium, and thus pressured broadcasters for jobs and other guarantees.[7]

Tension mounted. Broadcasters insisted that television was in no position to employ staff orchestras or match the pay scales of radio. At the same time, the networks ignored Petrillo's demands concerning FM broadcasting. As a result, in October 1945 Petrillo sent a telegram to broadcasters threatening a strike against the networks, which now numbered four since antitrust rulings had forced NBC to sell its Blue Network, which became the American Broadcasting Company (ABC). Petrillo demanded that broadcasters employ separate orchestras of equal size for AM and FM channels if they duplicated programs on the two media. Standby fees were not acceptable in lieu of the actual employment of musicians. To encourage an independent FM medium, Petrillo urged AFM locals to negotiate agreements of their own with independent FM stations.[8]

Broadcasters faced a dilemma. To hire double crews, as the union demanded, or offer musicians additional wages for duplicating their services would raise the cost of labor and perhaps encourage other workers, such as writers and technicians, to make similar demands. Yet the networks were ill prepared for a strike. Their most popular programs depended on live musical accompaniment; indeed, live music had become a trademark of network broadcasting, one of the things that distinguished it from local programming.

In the face of Petrillo's demand, therefore, the networks decided to halt FM broadcasting. The fact that the FCC had recently altered FM wavelength assignments influenced this decision, for the new assignments could more easily be achieved if all stations simultaneously shut down their FM operations while they modified their transmitters. Kesten explained the action of CBS in this matter to its affiliated stations in telegrams that sharply criticized the union. The networks, he said, "cannot assume the impossible burden which would result from the musicians' demands." Doing so would "seriously retard the development of FM broadcasting," added

Kesten, who hoped the FCC would find ways to protect the industry. But the FCC could do nothing. In a speech to broadcasters in Cleveland in October, FCC chairman Paul Porter reminded them of the limits of the commission's power. The commission "is in favor of duplicate programs," Porter said, "but Petrillo has overruled the FCC."[9]

The broadcasters' continuing hopes of help from the FCC were dashed in March 1946, when the commission released a report criticizing the programming policies of local stations. The report, *Public Service Responsibility of Broadcast Licensees*—nicknamed the Blue Book because of the color of its cover—called attention to the disparities between what broadcasters promised when applying for license renewals and what they actually did after the licenses were renewed. In accordance with FCC codes the broadcasters invariably promised to make time available for local programming employing local talent. A study of over eight hundred program logs, however, found that broadcasters generally broke those promises and filled the air instead with advertisements, recorded music, and network programming. Most stations, the report said, were "mere common carriers of program material piped in from outside the community." As a result, the average local station employed no musicians or actors. The report suggested that the FCC should no longer automatically renew broadcasting licenses but should instead compare the promises and performances of stations and act as the comparison dictated.[10]

Broadcasters and advertising agencies responded to the report with a barrage of criticism. *Broadcasting* described the report as "contrary to the precepts of the Constitution" and compared it with developments that "led the German and Italian people down a dismal road" to Nazism and fascism. Similarly, the head of a leading advertising agency, Lewis H. Avery, accused the FCC of seeking to impose "a diet of forced feeding on the American listening people." Justin Miller, head of the National Association of Broadcasters (NAB), suggested that station owners refuse to release program logs to the FCC to test the constitutionality of the agency's licensing powers.[11]

Such rhetoric gave the AFM little incentive to back away from its hard-fisted tactics. On the contrary, it affirmed the conviction of union leaders that broadcasters would never willingly share the profits of FM and television, and that wrestling concessions from them would therefore be difficult, perhaps even impossible. Moreover, by withholding musical services from FM and television, the union again opened itself to charges of Luddism and to criticism as well from some musicians in media centers. Yet how else could the union protect the interests of instrumentalists? Experience showed that only aggressive activity had any chance of succeeding.

WHILE MUSICIANS pondered these matters, Petrillo and the union became objects of new initiatives in Congress. In 1944 Petrillo had been a symbol of defiant labor, and a group of congressmen tried for the first time to pass legislation to curb him and his union. Republican Senator Arthur H. Vandenberg of Michigan had proposed an amendment to the Communications Act of 1934 to prohibit union interference with noncommercial and educational programs on radio. The Senate passed the amendment, but the House of Representatives adjourned without doing so.

In 1945 the Senate again passed the amendment, while the House authorized an investigation of "coercive practices in broadcasting." The Committee on Interstate and Foreign Commerce, chaired by Democrat Clarence F. Lea of California, conducted the investigation between February and May and concluded that Petrillo was indeed guilty of "abuses of power" that necessitated corrective legislation. As a result, in early 1946 George A. Dondero, a Republican from Vandenberg's home state of Michigan, introduced legislation in the House aimed at Petrillo. The bill called for much harsher measures than the Senate had contemplated in passing the Vandenberg amendment.

The "Anti-Petrillo Act," as the bill came to be known, made it illegal for employees in radio to use "intimidation" or "other means" to force broadcasters to hire persons "in excess of the number of employees needed" or to pay for services "which are not to be performed." Violators could be imprisoned for up to a year and fined up to $1,000. Clearly, the purpose of the legislation was to outlaw the union practice of demanding minimum-size and standby orchestras, a practice that was the source of many musicians' income.[12]

During the House debate over the bill, Democrats and Republicans alike described Petrillo in language reminiscent of that used in newspaper editorials during the early weeks of the recording ban. He was variously a "racketeer," a "power-grasping dictator," and a "big rat" who, as the occasion demanded, used "larceny," "embezzlement," or "extortion" to get his way. Lea and Dondero led the attack but found strong bipartisan support. Representative Lyle H. Boren, a Democrat from Oklahoma, called Petrillo a "despot who tramples upon the democratic principles under which the American people want to live." Harris Ellsworth, an Oregon Republican, agreed. "We have weeds in gardens and we have pests to bother animals and human beings," Ellsworth said, "and temporarily . . . we have Petrillo to bother the American broadcasting industry." Democrat Chet Holifield of California compared Petrillo with another unpopular labor leader, John L. Lewis, calling both men a "stench in the nostrils to legitimate organized labor unions."[13]

The debate made it clear that Petrillo's recent demands upon FM and television lay behind the bill. In fact, the report of the Lea committee contained a copy of the telegram Petrillo had sent to broadcasters prohibiting AM-FM duplication. Lea called the prohibition an "absurdity" that would force broadcasters to "needlessly duplicate" music. Under the prohibition, he explained, an AM radio station with a ninety-five-piece orchestra—a ridiculously inflated figure, considering that the average size of radio orchestras was probably six or eight pieces—would have to employ 190 musicians to broadcast music over an FM channel. Petrillo, warned Democratic Congressman L. Mendel Rivers of South Carolina, "has it in his power to kill once and for all frequency modulation. If he continues, television is dead and buried."[14]

The little support Petrillo had in the debate came from congressmen representing centers in which the AFM and its members had influence in local politics and labor councils. That resolute friend of labor, Vito Marcantonio of New York, one of two congressmen from the American Labor Party, called the attack on Petrillo a "smokescreen" designed "to prohibit the average American musician from getting some share of the enormous profits that come out of these [entertainment industry] monopolies." Democrats Benjamin J. Rabin and Emanuel Celler, also from New York, pointed out that musicians were trying desperately to fend off technological displacement. Rabin insisted that the real question was whether musicians would "get their share of the wealth that is created by these new machines," while Celler maintained that musicians who were receiving $3 million annually from the recording industry were taking "the place of live musicians who would receive for their work approximately $100 million." Celler added, "Musicians are getting a raw deal from canned music," and he urged his colleagues "to hold out a helping hand" and not "slash and smash" musicians with a "Draconian" law. Democrat Adolf J. Sabath from Chicago agreed. "Even musicians," he said, "have to eat."[15]

Such pleas fell on deaf ears. On February 21 the House passed the Lea bill by the lopsided margin of 222 to 43. When some senators suggested that the House bill differed too much from the Senate-approved Vandenberg bill, worried union officials breathed sighs of relief. But in the conference to reconcile the bills, Lea and other House conferees defended the House version, and the bill that emerged from the conference was that version virtually intact. Despite impassioned pleas from Marcantonio, the House approved the conference bill, as did the Senate, with barely a quorum present, forty-seven to three. Ten days later President Truman signed the act into law, making musicians another group of workers against whom Congress had enacted a specific law.[16]

Musicians responded to the new law with denunciations of their own. *Overture* suggested that Congress had become "hysterical" and "voted away" the rights of musicians. "Reactionary politicians," the editorialist explained, had gone "on a rampage" at the behest of broadcasters who had persuaded Congress to "paralyze" musicians because the broadcasters were "quite frankly worried about paying for FM and television." At the union's annual convention in St. Petersburg, Florida, Petrillo repeated the charge that the new law stemmed from the cozy relationship between Congress and radio. Congressman Lea himself was a close friend of NAB president Justin Miller, Petrillo noted, and that friendship had helped Miller rise in the ranks of the NAB. Petrillo similarly accused Congressman Eugene Cox of Georgia, who had called Petrillo a "racketeer" in the House debate, of accepting $25,000 from broadcasters in return for influencing FCC policy. He also reminded delegates that Democratic Senator Burton K. Wheeler owned a radio station in Spokane, Washington, and was a member of the NAB.[17]

In saying these things Petrillo had scratched the surface of a growing and problematic link between Congress and radio. Surveys of station ownership in trade journals help explain why many lawmakers supported the interests of broadcasters. In 1946 several House members, including Republicans Arthur Capper of Kansas, Harris Ellsworth of Oregon, Alvin E. O'Konski of Wisconsin, and John Phillips of California, as well as the wife of Democratic congressman Lyndon B. Johnson of Texas, owned radio stations. At least two senators besides Wheeler did also: Republicans Chan Gurney of South Dakota and William F. Knowland of California.[18]

Other business interests tied other members of Congress to broadcasters, though in less direct ways. Republican senator Homer E. Capehart of Indiana, for example, who had once owned a radio-manufacturing firm, had investments in a commercial record company. Personal friendships and kin relationships linked other congressmen besides Lea to radio. *Broadcasting* referred to Senator Vandenberg as a "lifelong friend and confidant" of the owner of two Michigan stations. Republican Robert A. Taft, then head of the Republican Party's policy committee in the Senate, had relatives in radio. Taft's cousin, Hulbert Taft, Jr., owned two Cincinnati stations and was president of Transit Radio Incorporated, a multimillion-dollar business that linked FM radio to public transit services. Another relative, David G. Taft, was director of a recently formed organization that advanced the interests of FM broadcasters. This pattern of relationships led union counsel Joseph Padway to tell the annual convention of the union in 1947 that the origins of the new antilabor laws lay in the "pressure of men such as RCA's David Sarnoff and CBS's Bill Paley."[19]

Passage of the Lea Act cannot be explained by such financial and personal relationships alone. Perhaps more significant was the context of the times. The year 1946 was one of the most tumultuous in the history of American labor. In January 1.2 million workers struck the automobile, electrical, and steel industries. During February the number of workdays lost to strikes nationwide totaled twenty-three million, approximately 3 percent of all work time. Nor was the unrest limited to industrial workers. Many teachers, public utility workers, and other service sector employees also walked off their jobs. By the time the Lea bill reached Congress, lawmakers at both the state and the federal level were far less tolerant of strike activity than they had been only a short time before. In fact, the Lea bill was but one of a growing number of antilabor initiatives undertaken in 1946. Only a veto by President Truman prevented the Case Act—which outlawed several longstanding trade union practices, including the right to strike without giving prior notice—from becoming law.

This was also a time when more and more Americans were worrying about the spread of left-wing ideologies at home as well as abroad. The deterioration of Soviet-American relations and the problematic future of capitalism and democracy in Eastern Europe and elsewhere raised questions about the relationship between militant trade unionism and the nation's way of life. Communists in fact dominated the leadership of some trade unions, especially at local levels. Leaders of various automotive, electrical, and maritime unions, to illustrate, endorsed communist ideology. The votes on the Lea Act reflected in part these growing concerns about communism and in part the emerging backlash against union activism and criticism of American industrial practices.

Many congressmen no doubt understood that musicians faced serious problems because of changing technologies and ways of doing business in the entertainment industries. But they also understood that some of the practices of Petrillo and his union adversely affected hundreds of businesses that contributed to the nation's economy. Indeed, musicians were not the only ones with stakes in recording and broadcasting. The livelihoods of thousands of technicians, salesmen, assembly line workers, and other employees outside of music depended on the radio and recording industries. Many lawmakers might well have concluded that for the sake of society at large, musicians and their union would have to adapt to new technology and to the new levels of productivity and production costs that the technology made possible, even though doing so reduced and centralized their job opportunities.

Most Americans, however, probably did not know or understand the significance of the Lea Act. For two decades musicians had maintained a considerable measure of control over the workplace in radio despite technological and organizational changes in the industry. The Lea committee acknowledged this fact but gave it a negative gloss. Because of the union's "coercive efforts," the committee found, "the industry has been forced to comply [with union demands] rather than suffer the penalizations that would follow."[20] By outlawing these "coercive" practices, the Lea Act more than punished Petrillo; it fundamentally altered the balance of power between musicians and their employers.

OPPONENTS OF the Lea Act questioned its constitutionality from the beginning, arguing in both House and Senate that it violated constitutional guarantees of free speech and equal protection of the laws and prohibitions against involuntary servitude. By prohibiting musicians from using "intimidation" or "other means" to accomplish their goals, critics maintained, the act violated First Amendment protections of the right to strike and picket and to speak freely. They insisted too that by singling out radio employees, the act breached Fifth Amendment guarantees of due process and equal protection of the law. Finally, they suggested that the act violated Thirteenth Amendment safeguards against involuntary servitude, since it apparently limited the right of musicians to refuse to work.[21]

At the suggestion of Joseph Padway, union leaders set in motion a plan to test the constitutionality of the act. On May 28, as president of the Chicago local, Petrillo asked radio station WAAF to add three musicians to its workforce, promising a strike by station musicians if the request was refused. (Actually the station employed no musicians but did employ three members of the AFM as librarians.) When the owner refused, Petrillo called the union members out on strike and placed a token picket line of one man in front of the station. Explaining his actions to reporters, Petrillo admitted to violating the Lea Act but said he was ready "to face the music." At the union convention a week later he told delegates, "I am now waiting for a marshal of the United States to arrest me."[22]

After the FBI and the Department of Justice reviewed the case, U.S. attorney J. Albert Woll filed suit against Petrillo, maintaining that he had violated the law by attempting to coerce a licensed radio broadcaster into hiring unnecessary employees. Petrillo's action, Woll said, amounted to racketeering. After an initial hearing, Petrillo posted bail and Padway petitioned the court to dismiss the case on the grounds that the Lea Act was unconstitutional.[23]

On December 2, 1946, Judge Walter J. La Buy elated musicians by endorsing Padway's argument and dismissing the charges against Petrillo. In an eight-page opinion La Buy ruled that the Lea Act violated the First, Fifth, and Thirteenth Amendments. He also ruled that Petrillo had not demanded that WAAF hire musicians "in excess" of the number needed because "there is no means, or guide, or standard by which the defendant may know 'the number of employees needed.'" Petrillo was not present when La Buy read his decision, but he quickly called a press conference and praised the judge for "upholding the constitution."[24]

Petrillo's celebration was short-lived. The Department of Justice promptly appealed La Buy's ruling, and on June 23, 1947, the Supreme Court upheld the appeal and the constitutionality of the Lea Act and ordered that Petrillo be tried on the original charges. Writing for the high court, Justice Hugo Black denied that the Lea Act unfairly singled out employees in radio, and he rejected the argument that the phrase "more employees than needed" was unconstitutionally vague. Black abstained from ruling on whether the law violated the First and Thirteenth Amendments, but he found that "the statute on its face is not in conflict with the First Amendment." He remanded the case to La Buy for trial.[25]

News of Black's decision was one of two shocks the AFM received on the same day. The other was news that Congress had passed a new and far more restrictive labor law aimed not just at musicians but at the trade union movement itself. The Labor-Management Relations Act of 1947 represented the culmination of a long campaign to amend the Wagner Act, the basic piece of New Deal labor legislation and the law most responsible for the rise of mass unionism in the 1930s and 1940s. The new law bore the names of the men who introduced it, Senator Robert A. Taft of Ohio and Representative Fred A. Hartley, Jr., of New Jersey, both Republicans, but it was the offspring of a coalition of conservative political and business leaders that dominated Congress after the elections of 1946. The National Association of Manufacturers, among many other economically powerful and politically influential groups, lobbied for the bill, and the Republican National Committee paid corporate lawyers to draft it. President Truman described the Taft-Hartley Act as "bad for labor, bad for management, and bad for the country," but the Republican-controlled Congress overrode his veto of it.

By defining many traditional, and traditionally effective, union tactics as "unfair labor practices," the Taft-Hartley Act was a direct assault on trade unionism itself and thus on the ability of laboring people to protect their interests through collective action. Among its provisions, perhaps the most important for the trade union movement generally was one that allowed states to pass "right to work laws" banning closed shops (all-union workforces). The new law further restricted organized labor, including the AFM, by outlawing sympathy strikes and secondary boycotts. Sympathy from other unions in entertainment industries, especially the International Alliance of Theatrical and Stage Employees (IATSE), had helped striking musicians on several occasions, and musicians had returned the favor. The unambiguous provision in the Taft-Hartley Act against secondary boycotts clearly prevented the AFM from pulling orchestras from network programs in order to assist musicians in network-affiliated stations. As these examples suggest, the Taft-Hartley Act significantly increased managerial control over industrial life by making it far more difficult for workers to challenge their employers.

Measures in the new law pertaining to hiring policies and employer welfare funds presented special challenges to musicians. Section 8(b), which prohibited unions from forcing employers to pay for services not performed, not only outlawed a longstanding practice of the AFM; it also gave broadcasters new grounds for refusing union demands for minimum-size crews and standby fees. In fact, when a puzzled senator asked for a definition of this provision, Taft referred to the AFM practice of demanding that broadcasters employ more musicians than they wanted or needed. Another provision of the law, which seemed even more hostile to musicians, prohibited workers from forcing an employer to "pay or deliver . . . any money or other thing of value to any representative of any of his employees who are employed in an industry affecting commerce."[26] This provision raised serious questions about the legality of the Record and Transcription Fund, since the fund had been "forced" from employers by collective bargaining.

Legally and politically, musicians had never been more vulnerable. In a span of fourteen months Congress and the courts had dealt them two severe blows. The federal government had aligned itself with the interests of employers and seriously undermined the ability of musicians to bargain collectively. The musicians' hard-won employment fund was threatened and their bargaining power in radio in shambles. Employment patterns soon showed the results of the changes. Backed by the Lea and Taft-Hartley acts, network affiliates downsized or eliminated orchestras as existing contracts expired. Between May and November about twenty radio stations discharged more than 140 full-time instrumentalists.

Ironically, as their power and job opportunities declined, growing numbers of musicians were joining the AFM. From 1944 to 1948, a period during which unionization among all skilled workers in the nation rose by about 6 percent, AFM membership jumped by 58 percent, from 147,000 to 232,000. The return of servicemen to civilian life cannot explain this anomaly. The unique pattern is probably attributable to the union's successes and Petrillo's high profile during the 1940s. The conflict and controversy that surrounded those successes, and that surrounded Petrillo himself, made Petrillo as well known as any other labor leader in the nation, including even John L. Lewis. Even though journalists and public officials often portrayed him as nothing more than a tin-pot dictator, Petrillo was to thousands of marginalized musicians a champion they identified with as working people. In 1943 and 1944, when Petrillo and his union wrested royalties from leading record and transcription firms, many of these musicians apparently concluded that union cards provided them the best hope they had, not only for jobs but also for the sense of control over their own lives that promised the fulfillment or self-worth they craved. But the extraordinary growth of the union could also be attributed to the public image of popular big bands and the exciting lifestyles they seemed to represent. Contrary to the union's own predictions about the consequences of mechanization, large numbers of young people continued to learn to play musical instruments, apparently in hopes of entering the small, elite, and glamorous groups of musicians who dominated public images of working musicians. Closed-shop hiring policies in clubs, restaurants, and radio stations no doubt also encouraged star-struck youth to join the AFM.

Whatever the explanation, the burgeoning growth in union membership affected large and small locals alike. Between 1944 and 1948, membership in New York Local 802 rose from 22,000 to 31,500, while that in Toledo Local 15 increased from 413 to 582; Milwaukee Local 8 grew from

1,579 to 2,268. During the same period some locals more than doubled their membership. In Sacramento, for example, membership jumped from 319 to 771, while in Cincinnati that of the union of black musicians grew from 53 to 124. The swelling size of the AFM undoubtedly emboldened Petrillo in his dealings with management, but it must also have reminded him of the pressing need to augment employment opportunities.[27]

PETRILLO HAD LED employers to expect an all-out campaign by the union to protect the interests of musicians. At the 1946 convention of the AFM, he talked of a nationwide radio strike if the courts upheld the Lea Act, and he promised to halt all recording if record companies stopped paying into the record-royalty fund. According to the *New York Times*, delegates rose to their feet to cheer this resoluteness, which Petrillo reaffirmed the next year by backing a proposal to give union leaders the power to initiate a second nationwide recording ban.[28]

Petrillo's determination to fight became more apparent shortly after the 1947 convention, when he and other union leaders met network representatives to renegotiate industrywide contracts, which were scheduled to expire on January 31, 1948. Months before, Petrillo had told broadcasters that mounting problems in the industry might prevent renewal of the contracts. He clearly hoped that a warning would cause industry leaders to make concessions on FM and television and to pressure their affiliates to maintain staff orchestras large enough to appease the union. The broadcasters, however, brushed off the warning and agreed only that negotiations for new contracts should begin in due course. This prompted the union to raise the stakes in what became a war of nerves. When industry leaders left Chicago following the initial negotiations, the executive board gave Petrillo power to decide whether negotiations were satisfactory or whether the existing contracts should be allowed to expire.[29]

Petrillo also threatened to strike the recording industry. In that industry the problem was the record-royalty fund and how to save it. After three days of wrestling with the problem, union leaders decided in October that there was no alternative to a second ban on recording. The board therefore approved a motion that union members "cease making records and transcriptions on expiration of [existing] contracts." Union musicians, the board announced, would "never again" make recordings, since "ultimately the making of same will destroy the employment opportunities of musicians." Nine days later Petrillo informed industry leaders of the decision. The contract between the industry and the musicians "will not be renewed," he told them. "On and after January 1, 1948, members of the American Federation of Musicians will no longer perform [for record manufacturers]." It is "our declared intention," he said, "permanently and completely, to abandon that type of employment."[30]

The AFM had become engaged in what appeared to be a two-front war against broadcasters and recorders but was in fact a single fight against a unified enemy. Mutual interests accounted for that unity. Without new records, broadcasters lost an essential source of programming material, and without radio, recorders lost the best and cheapest way of advertising their products. The ties uniting the two industries became closer as entrepreneurs in the one expanded into the other and into allied economic activities. In the postwar years radio and recording firms had become basic components of interlocking interests. RCA, to illustrate, not only owned NBC but also, through other subsidiaries, produced phonographs, television sets, radios, and other entertainment-related products. A musicians' strike would adversely affect all of these activities, and not surprisingly, all of the industries of which they were parts banded together to resist the union.

Cooperation between employers reached new heights in late 1947, when industry leaders organized the All-Industry Music Committee (AIMC), the most formidable new organization musicians had faced since the rise of the NAB. The structure of the AIMC reflected the unity of employers as well as their determination to thwart the union. An executive committee directed overall strategy against the union, while separate subcommittees dealt with legal issues and public relations concerns. Representatives of the NAB, the networks (including their FM and television subsidiaries), record and transcription companies, and radio manufacturers sat on the executive committee as well as all subcommittees. To finance the organization, member companies contributed according to their gross earnings, which meant that major broadcasting and recording companies footed the bill.[31]

The strategy of the AIMC became clear in December, when the executive committee gathered in New York to select heads of the legal and public relations subcommittees. After the gathering adjourned, NAB executive A. D. Jess Willard, whom *Variety* called the "flywheel" of the group, explained the purpose of the organization. Petrillo had inflicted "grave injustices" on music businesses, Willard noted, and the AIMC intended to "acquaint the public with the facts" about him and his union and how they threatened the future of recording and radio. The AIMC was necessary, he said, because of the common threat to all segments of the industry. Employers "had exchanged ideas and information," Willard added, "in order that . . . no one group goes off on a tangent"—and, he might have added, to ensure employer solidarity in the looming conflict with musicians.[32] If it worked, industry might well emerge from the conflict in firm control of its workforce.

Manufacturers again stockpiled recordings. In a study of the industry at the time, Russell Sanjek, a former vice president of Broadcast Music Incorporated (BMI), noted that as soon as recorders realized that musicians were serious about a new boycott, they speeded up production. "Columbia and Victor," Sanjek said, "invested two million dollars in a down-to-the-wire frenzy of record cutting, producing 2,000 masters at an average cost of $1,000 each." Sanjek explained that the speed-ups were due partly to the fact that the union had stipulated in 1944 that in the event of another recording ban, existing contracts with the union would be invalidated. Recorders would thereby lose exclusive control over the services of star performers, which could eventuate in "a bidding war for talent."[33]

These developments worried AFM officials, but union leaders had more specific incentives to settle their differences with employers. Unlike the first recording ban, the new one would also affect musicians in radio, and therefore it promised greater hardship for rank-and-file musicians and thus for union solidarity itself. Most musicians made recordings to supplement their income from other sources, but hundreds of instrumentalists depended on their jobs in radio for their basic income. A lengthy strike against radio and recording, then, promised to drain union strike funds and foster internecine strife. The union was therefore anxious to keep the door open on negotiations with radio.

With these concerns in mind, Petrillo agreed to meet network representatives in New York in November. To map his strategy, he called together the executive board and representatives from the New York and Los Angeles locals. "He told us," Local 47 representative Phil Fischer later said of the meeting, "that where heretofore we used to come in and make 'demands,' we could now only 'negotiate,' with the cards stacked against us in favor of the employers. He told us too that . . . one ill-advised remark might immediately involve us in a violation of [the Lea or Taft-Hartley Act] and throw a monkey-wrench into the negotiations."[34]

At the negotiations, which began on November 19, Petrillo acknowledged that the new laws made new patterns of bargaining necessary. Instead of making demands, as he had previously done, Petrillo asked the network representatives to present their vision of the future of musicians in broadcasting. "Mr. Petrillo made the opening remarks," Fischer recalled, "[that] times have changed and with new laws now on the books, everything is in favor of the employers." He "told them the truth, that you gentlemen know we are not going to 'demand,' we want to know what you want." *Broadcasting* described Petrillo's tactic as a "reversal of his usual [practice] of starting off with exorbitant demands" and suggested that it "caught the nets off guard." Nevertheless, network representatives had a "bill of particulars," which they promptly presented to Petrillo.[35]

The "bill" focused first on FM and the intent to duplicate live musical performances on AM and FM with no added pay for musicians. The duplication, the networks noted, required no additional labor and meant no additional cost to sponsors.[36] The networks intended too to use their radio orchestras for television programming, also with little or no extra pay for musicians. The fledgling television industry, they explained, could not yet afford full-time musicians of its own, or even part-time workers paid at radio wage rates. Broadcasters also proposed to use recorded motion-picture music in television programming and concluded by suggesting that the future of musicians in radio hinged on union policies regarding FM and television. The AFM, they reminded Petrillo, had long burdened radio with unnecessary costs. The networks, they also insisted, had no control over the hiring policies of their affiliates and "could not be of much assistance" in protecting union jobs outside media centers.[37]

In short, the networks would hire no more musicians than they needed on FM or television, and they would not be intimidated by threats of a strike. This stance was all the more ominous because Petrillo and other union leaders had already concluded that a strike would hurt musicians more than broadcasters. "We knew that by this time," Fischer said. "The chain companies were prepared for a nation-wide strike in radio. The fact that they spent at least $200,000 in recording themes and bridges for all the radio shows was conclusive proof." Petrillo himself "did not think that a strike would be of any benefit"; and in response to the network proposals, he simply agreed to study them and offer counterproposals in early December. Speaking to reporters after the negotiations, Petrillo admitted that musicians were "worried"; the unknown consequences of FM and television as well as the new labor laws made it more difficult to know how or what to negotiate. "It's not so easy now," he said plaintively.[38]

Petrillo's response to the industry proposals revealed just how much the new laws had changed his bargaining power. He made no demands. "We would like to have an increase in the number of staff musicians in New

York, Chicago, and Los Angeles," Petrillo told network representatives, and in affiliated stations as well. He urged the networks to encourage their affiliates to return employment to the levels that had existed before passage of the Lea and the Taft-Hartley acts. He also urged them to hire only union members as "pancake turners" (disc jockeys). The last and rather unexpected request stemmed from the fact that Petrillo and the union had long argued that disc jockeys rendered a musical service and should therefore belong to the AFM. The request, however, seemed to trivialize Petrillo's final proposal: that network musicians wanted "a substantial increase in wages."[39]

These exchanges occurred while the union and the networks awaited the outcome of Petrillo's trial in the case involving station WAAF in Chicago. Union officials anticipated a favorable decision in the trial, because Judge La Buy had sided with Petrillo in the original hearing. His ruling, they hoped, would improve their bargaining position, which was one reason Petrillo played for time in the negotiations just discussed. Petrillo appeared before La Buy a second time, and instead of challenging the constitutionality of the Lea Act, he pleaded not guilty to the charges of forcing WAAF to hire unneeded musicians and waived his right to a jury trial. The tactic paid off; on January 14, 1948, La Buy exonerated Petrillo.[40]

The ruling kept Petrillo out of jail and discouraged prosecutions of other union leaders under the Lea Act. But it did nothing to deter radio stations from discharging staff musicians. Provisions of the Taft-Hartley Act prohibiting secondary boycotts prevented Petrillo from pulling musicians off network programs to safeguard jobs in affiliate stations, as he had heretofore done. Within six weeks of La Buy's ruling, in fact, four stations eliminated or downsized their radio orchestras, costing twenty musicians their jobs. WAGA in Atlanta discharged three musicians, while WKBW in Buffalo discharged eight, and two stations in East St. Louis, WTMV and WPEN, together discharged nine.[41]

BY THIS TIME Petrillo's struggle with employers was again the subject of congressional attention. Fred A. Hartley, Jr., of New Jersey, chairman of the House Committee on Education and Labor and co-author of the Taft-Hartley Act, appointed a new subcommittee to investigate the mounting criticisms he and others in Congress had heard of Petrillo and his union. Chaired by Republican Carroll D. Kearns of Pennsylvania, the subcommittee held hearings during the summer of 1947. On July 7 and 8 Petrillo testified at the hearings, fielding criticisms from Kearns as well as Republican representatives Richard M. Nixon of California and O. C. Fisher of Texas, and Democrat Graham A. Barden of North Carolina. The hearings produced no evidence that Petrillo or the union had broken any law, but they set the stage for a barrage of attacks upon him and his dealings with broadcasters. Petrillo, the committee found, wielded the kind of "tyrannical power" that "should not be countenanced nor tolerated in a Free Republic." He and his union had "held back the technological development of radio" and threatened "to block the . . . development of television." The recording ban that he was now talking of would "close down over 500 recording companies" and "throw out of employment thousands of people." Congress should therefore pass legislation curtailing the "monopolistic practices of labor unions which are injurious to the public interest." Among the practices to be curtailed was that of calling industrywide strikes "such as is threatened by Petrillo in the Recording industry."[42]

As a result of the subcommittee report, Hartley scheduled further hearings before the full committee. Broadcasters and other employers of musicians used the hearings to advance their own interests against those of Petrillo and the union. The AIMC spearheaded an effort to line up witnesses and coordinate their testimony. Executives of major radio and record companies and heads of several employer associations testified on behalf of the industry. NAB president Justin Miller, to illustrate the thrust of their testimony, admitted that industrialists cooperated closely when dealing with Petrillo's union. "Ordinarily, I would not be speaking for the recording companies, the transcription companies, the manufacturers or any of these other groups," Miller testified, but on "this particular problem we have all gotten together." As evidence of Petrillo's "dictatorial" powers, Miller pointed to an interesting fact: fifty-five thousand musicians, roughly a fourth of all AFM members, lived in three media centers, yet at union conventions the locals in these centers cast only 30 of 1,445 votes. This "undemocratic" structure, Miller maintained, accounted for Petrillo's inordinate control over music services. Miller saved his harshest criticism for Petrillo's stance concerning FM radio and television. Petrillo's refusal to permit musicians to work in these media, he insisted, again showed the union leader's "opposition to modern technology."[43]

The highlight of the hearings came on January 21, when Petrillo himself testified. After a week of testimony against the labor leader, committee members had scores of questions for him. "My testimony," Petrillo said later, "dealt with such diverse subjects as democracy in the AFM, foreign broadcasts, unemployment of musicians, contracts with the motion picture industry, television and FM, amateur orchestras, and the recording ban." For more than six hours Petrillo answered questions, once again taking advantage of the spotlight under which his critics had placed him. Asked why people thought of him as a czar or a dictator, he responded by blaming the NAB's "hold" on "the distribution of communication and news." He and his union, he explained, were "up against one of the greatest propaganda machines that the city of Washington DC, was ever faced with." Using its control of "some 400 newspapers [and] every radio station in this country," he said, the broadcast industry shaped the news for its own purposes and slandered whomever it pleased. "No one was ever more vilified than I have been in the press," Petrillo contended. "If they would spend half of the money that they spend on cartoons vilifying me as the president of this organization, if they would give it to the musicians, we would all be happy."[44] Petrillo ridiculed charges that he or his union opposed technological innovation. "I don't think anyone in this country," he scoffed, "is big enough to stop progress." His union had simply tried to prevent employers from using technology to the detriment of musicians. "We are being destroyed," he told the committee, and "are trying to protect ourselves in the best way we know how." Musicians were "ready and willing" to provide services for FM and television, but not on whatever terms their employers offered, and they would return to recording studios when Congress found a way to protect the Record and Transcription Fund.[45]

Petrillo's performance was impressive. *Variety* thought his "savvy" and "showmanship" had "provided plenty of entertainment for the committee," while the *Washington Post* called his performance "such that no union member could complain." *Broadcasting* conceded that he had been "difficult to pin down" and worried that he had effectively countered employers' attempts to "rattle the chandeliers with tales of suffering." Even Hartley, who remained convinced of the need for greater restrictions on the power of labor leaders, told reporters, "Mr. Petrillo is a good witness. He's disarming by his absolute frankness." Broadcasters and recorders, Hartley added, should have presented a better account of their problems with the union.[46]

A day after Petrillo testified, the AFM's new legal counsel, Milton Diamond, delivered a more informative and eloquent defense of union practices. Diamond had previously worked for Decca Records, which he helped organize and for which he had been legal counsel and associate chairman of the board of directors. Petrillo selected Diamond to succeed

Gid-Dap!

The press seemed to enjoy portraying Petrillo as a tyrant who opposed progress and ignored the public interest. This Shoemaker cartoon appeared in the *Chicago Daily News,* October 23, 1947.

the deceased Joseph Padway as part of a new public relations campaign. Diamond was a respected, mild-mannered expert on the financial and legal structures of the entertainment industry, and his public statements on the union's behalf were all the more effective for his tact. With candor and detail Diamond refuted the charges made in the hearings against the union, and in doing so he undermined the notion that Petrillo had acted against the public interest.[47]

More important, Diamond put the problems of musicians in their social and economic context. The basic question for Congress, he explained, was not how to punish the AFM or its leader but "how to deal with technological displacement of human labor." New methods of recording and broadcasting, he noted, "have displaced, or have the potential of displacing, all but a few of the thousands of musicians who have studied and trained from childhood that their bread might be won by the practice of their profession." Musicians understand "that these wondrous accomplishments have implicit in them the seed of the destruction of musicianship," but they do not oppose progress. Petrillo was not "a modern-day Canute," Diamond said, "peremptorily bidding the tide of scientific progress to halt and recede." He was instead a union leader committed to the well-being of workers confronting problems of technological change.[48]

The effect of Diamond's testimony is unclear, but the Hartley committee made no legislative recommendations. What could the committee have done? Petrillo had violated no law, and his position within the union was secure. "There is no question," the Kearns subcommittee had concluded, "that Mr. Petrillo has the backing of most of the members of the American Federation of Musicians."[49] The structure of the union and the powers of the president had not changed significantly in half a century. What had changed were employment patterns and opportunities for musicians. It was largely the response of musicians to these changes that brought them and their union to the attention of Congress.

The things said in these public hearing widened the ideological as well as the rhetorical gap separating Petrillo and his critics. Petrillo considered the statements of political and industry leaders insulting. His critics had portrayed him as a backward-looking tyrant, a selfish agitator unmindful of the public good. In defending himself Petrillo presented an interpretation of business history that similarly offended his adversaries. His interpretation stressed themes of monopoly, heartless destruction of jobs, and other evil consequences of concentrated wealth; it also suggested that only aggressive collective action could protect workers from those conse-

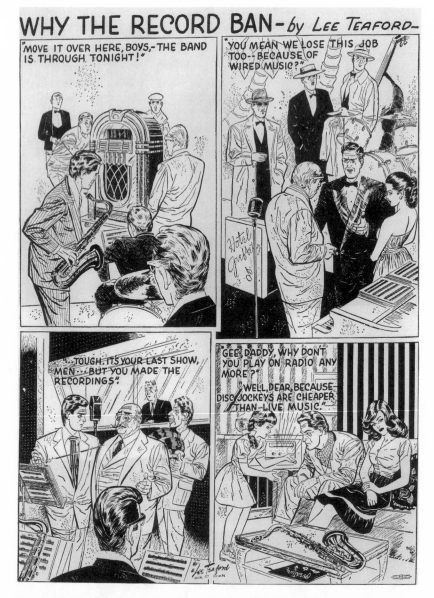

The AFM labored to answer attacks on Petrillo and justify the second recording ban. This graphic explanation appeared in the *International Musician* of January 1948.

quences. Whatever the merits of these views, they did not seriously threaten employers' control of economic life.[50]

AFTER THE HEARINGS Petrillo and other union officials resumed contract negotiations. On January 26, only five days before the contracts were due to expire, they began a series of talks with network representatives in a final effort to reach an agreement. For several days proposals and counterproposals crossed the negotiating table. Petrillo made the first concession. He agreed to extend the current contracts for sixty days and to accept AM-FM duplication during that time if broadcasters charged advertisers no additional fees for the double exposure. He also agreed to quit pressuring the networks to get their affiliates to hire more musicians. These concessions reflected the union's weakened position. Petrillo surely knew that anything he conceded would be difficult to regain. The green light he gave to FM duplication meant that eight network stations began sending live musical programming to more than 225 outlets, and as the popularity of FM grew, consumers, employers, and public officials would resent any attempt to restrict the medium. Similarly, Petrillo's promise to quit pressuring affiliates through the networks was irreversible. The networks had always resisted the pressure, and the new labor legislation made it illegal.[51]

Industry leaders made the most of the contract extension, using it to work out a new negotiating strategy. According to *Variety,* NAB executives met on February 9 and resolved to "hold the line" against Petrillo, noting that it was "more important than ever" that they do so. Thus, as the resumption of negotiations neared, industry leaders were confident they could defeat Petrillo. "The word sifting through to [networks]," reported *Variety* in early March, "is that Petrillo is now anxious to get the whole affair settled as soon as possible."[52]

When negotiations resumed on March 8, Petrillo suggested that the first order of business be contracts for network orchestras. The union, he said, wanted the networks to increase the number and the wages of their musicians. Network representatives responded that any agreement concerning radio musicians depended on assurances that the musicians would perform on television. They also noted that the networks intended to reduce their musical staffs, and they held the line on wages. The jobs and income of staff musicians, in other words, hinged on new concessions from the union. When Petrillo demurred, the press predicted a strike against radio.

Within a few days, however, Petrillo retreated. When he told negotiators for the networks they would have to "fiddle along on tele" until musicians got what they wanted from radio, the negotiators, according to *Variety,* "arose and started to walk out in a body." At that point "Petrillo knew the jig was up and threw in the towel." The new three-year contracts he agreed to guaranteed only that network stations would maintain current levels of employment. In return, the union agreed that the networks could duplicate AM programs on FM and use AM orchestras on television for "reasonable" additional fees (soon set at 66 percent of radio wages). The contracts provided no wage increases.[53]

Petrillo's surrender not only revealed the new realities in the working world of musicians; it also showed how dependent on network employment musicians had become. More than 80 percent of the income of musicians in radio now came from fifteen network stations. To strike those stations would cause serious hardship for musicians and would cut off much of the union's own income from dues. Petrillo's action thus signaled a major shift in power relations in the industry. The new labor laws gave industry leaders almost complete control of broadcasting and recording technology, and as their control increased, the say-so of musicians over their own work declined.

UNION LEADERS hoped to balance these setbacks in radio with gains in recording. They had reason to be optimistic. Since no instrumentalists worked full-time in recording, the union could afford to maintain the recording ban. In addition, for hundreds of locals the Record and Transcription Fund had become a symbol of pride as well as a vital source of revenue. Support for the recording ban was thus strong. More important, growing numbers of record manufacturers were willing to settle on union terms and continue payments into the Record and Transcription Fund if provisions of the Taft-Hartley Act prohibiting such payments could be circumvented.

By the summer of 1948 union leaders had ample reasons of their own to settle the dispute. If the ban continued much longer, income from the record-royalty fund would disappear. The contracts requiring payments into the fund would expire on July 1, after which the status of the fund itself would become problematic. Furthermore, recording musicians had shown that union solidarity had its limits. Rumors circulated in media centers that record companies were offering musicians long-term employment to break union ranks. Some musicians were apparently contacting those companies and offering to ignore the ban for the right price, even though to do so meant certain expulsion from the union. Moreover, in-

Petrillo and radio network executives (from left, Mark Woods, American Broadcasting Company; Robert Sweezey, Mutual Broadcasting System; Joseph Ream, Columbia Broadcasting System; and Frank Mullen, National Broadcasting Company) enjoy a light moment after signing a three-year contract in March 1948. The trumpet was a gift from Mullen, which Petrillo agreed to play on the first live-music television program broadcast under the new accord. In December Petrillo finally lifted the ban on recordings. *(AP/Wide World Photos)*

creasing numbers of union musicians were making bootleg recordings. Saxophonist Lenny Atkins, who worked for CBS radio during the ban, recalled numerous recordings made in Mexico by AFM members. "Some musicians were taken out of town, to Tijuana," Atkins recalled. "I was shocked at the loyal musicians who went down there. They were hurting the cause."[54]

This disloyalty reflected the fact that some instrumentalists had grown weary of the ban. Musicians in media centers with successful careers in radio, recording, and film studios no doubt resented the periodic disruptions of business. They may have recognized the larger problems of the union, but for them the disruptions meant lost opportunities and income.

Union-imposed quotas, pulling remote broadcasts, and two recording bans had only made the adjustment to new business conditions more difficult for these musicians. Trumpeter Bob Fleming at Walt Disney Studios thought that by 1948 Petrillo had become "obnoxious." "I thought the union was overdoing it," Fleming said. Staff guitarist Roc Hilman at radio station KFI in Los Angeles agreed: "I was getting annoyed. I didn't go for it, the whole idea." Will Brady of the Kay Kyser Orchestra felt likewise: "The things [Petrillo] was doing with [the union] I didn't like at all." Brady and others nevertheless adhered to union policy. After all, as Brady explained, "Petrillo was pretty strong."[55]

There were other problems too. In the spring of 1948 the AIMC was considering the possibility of filing charges of unfair labor practices against the union on the grounds that the recording ban amounted to a strike against radio and was therefore a secondary boycott as defined by the Taft-Hartley Act. On May 13, attorneys for three transcription companies made the possibility a reality, arguing in separate complaints to the National Labor Relations Board (NLRB) that the union ban had forced transcribers "to cease doing business with . . . the proprietors of approximately 560 radio stations." The complaints asked for an injunction against the ban, according to provisions of the Wagner Act.[56]

Congress gave union leaders another reason to end the ban when it shelved legislation that would have allowed the union to collect record royalties from the jukebox industry. In early April a House subcommittee on patents and copyrights had recommended eliminating a clause in the 1909 copyright law prohibiting the collection of performance royalties from coin-operated music machines. (The clause had been written for the benefit of player-piano companies.)[57] Why the recommendation was not acted upon is unclear, but the lobbying efforts of the new trade association, the Music Operators of America, probably played an important role. Well-known crime reporter Lester Velie suggested that big-city mobs with stakes in jukebox businesses and influence in Congress were responsible. Whatever the reason, the failure to amend the copyright law dashed union hopes for a legislative resolution of the struggle over record royalties.[58]

This litany of setbacks, concessions, and anxieties made the annual convention of the union, at Asbury Park, New Jersey, in June 1948, one of the gloomiest in the organization's history. Clearly disheartened, Petrillo reviewed the preceding year without his customary bluster and bravado. "The cards were stacked against us," he said of his recent capitulation to the networks. "We had $26 million in wages tied up in radio and we were afraid that if we held out too long we might lose that." He apologized to small locals for being unable to safeguard jobs in local stations, explaining that the new labor laws made that impossible. Because of complaints to the NLRB, he added, he might also have to lift the recording ban. In an especially telling remark he admitted the union's vulnerability: "Industry is now running the show."[59]

But all was not lost. As the NLRB considered the complaints against the union and record supplies and sales dwindled, the chances of a favorable settlement with recorders improved. Negotiations continued from September to November, and it became increasingly clear that growing numbers of recorders would agree to continuing payments into a union employment fund. Eventually Milton Diamond and industry representatives hammered out an innovative five-year agreement that got around the legal obstacles raised by the Taft-Hartley Act.

The agreement involved a trusteeship arrangement that circumvented those obstacles by granting the power to collect and spend monies in the fund to a trustee acceptable to both the industry and the union. The trustee would collect from record companies a percentage of the price of each record sold and would in turn see that the revenue thus collected went to finance musical concerts for the public. The royalty payments still ranged from 1 to 2.5 percent of the price of records, transcription companies continued to contribute 3 percent of their gross revenues from leasing records, and AFM locals continued to benefit from the fund. This arrangement received sanction from the federal government in early December, when the NLRB ruled that the boycott of transcription companies did not violate the Taft-Hartley Act, and both the secretary of labor and the attorney general ruled that the trusteeship agreement did not conflict with Taft-Hartley.[60] The employment fund was thus secure.

The recording ban ended immediately. On December 14, 1948, Petrillo and representatives from eleven record companies signed the trusteeship agreement, which set up the Music Performance Trust Fund. Within three months nearly two hundred companies had endorsed the agreement, and Philadelphia attorney Samuel Rosenbaum had become trustee of the fund. The former radio station owner who represented affiliates in disputes with the union in the late 1930s and early 1940s had since become a friend and confidant of Petrillo and thus stepped into his new role with the union's blessing.[61]

As he customarily did, Petrillo hailed the settlement as a triumph for the labor movement as well as for instrumentalists. The union had indeed accomplished something meaningful. The settlement saved the royalty fund, thus preserving an important source of income for thousands of musicians and hundreds of union locals. In one year, from the middle of 1949 to the middle of 1950, the new fund provided more than $1.3 million to finance musical concerts, about the same amount the original fund had generated in 1945. In doing so it financed nearly eighteen thousand performances at veterans' hospitals, symphony halls, and other locations. In this modest but instructive way the AFM showed how labor might be compensated for the effects of technological change. Musicians, then, stood with workers in the automotive, trucking, mining, and other industries who had responded in similarly innovative ways to new circumstances brought on by technological innovation. Among these responses were early retirement plans, increased severance pay, shorter workweeks, voluntary retraining programs, job transfer policies, and assorted fringe benefits tied to specific forms of automation or other forms of technological change.

Securing the Music Performance Trust Fund, however, was a single and limited victory in a period of significant setbacks for musicians and their union. It maintained a benefit won four years earlier and in doing so protected an important source of occasional income for large numbers of musicians. But it was small compensation for the losses in radio or the blighted prospects in FM and television. These new media, musicians now realized, would create few jobs for them in the immediate future. More important, government-business-labor relations had evolved in ways that practically destroyed their union's ability to protect their interests. The collective effects of these changes were soon apparent. Between 1946 and 1950 musicians lost more than five hundred full-time jobs in radio stations affiliated with the networks, and in those years the annual earnings of staff musicians in the industry dropped from $12 million to $10 million. Simultaneously, earnings from single-engagement commercial broadcasting fell from more than $8 million to less than $5 million.[62] And this was but a portent of things to come.

Conclusion

THIS STUDY HAS EXPLORED aspects of the history of American musicians from the 1890s to 1950. It has focused on how the advent of new technologies in rapidly succeeding generations of greater complexity, higher quality, and increased productivity in several mass-entertainment industries gave rise to gigantic vertically integrated, capital-intensive, geographically centralized business enterprises. It has focused too on how those technologically driven developments transformed the working world, and thus the social and economic well-being, of musicians. More specifically, the study has tried to show that the introduction of sound movies, network radio, and high-fidelity recordings turned diffused, labor-intensive job markets and workplaces into more centralized and mechanized ones; and it has tried to describe and assess the social and economic costs of this transformation for working musicians. These final pages offer a chance to reflect on these changes and on how the musicians' experience of them illuminates points at which labor history and the history of business, technology, and society intersect. Specifically, they speak to the relationship between work and technology, the response and counterresponse of labor and management to technological change, and the varying roles of government and the market as ultimate arbiters in regulating the consequences of that change. This is also an opportunity to reflect on the implications of continuing technological change for American workers.

THE STORY OF musicians and the sound revolution began in an age before recorded music became a serious threat to the careers of instrumentalists playing before live audiences. In the late nineteenth and early twentieth centuries the demand for skilled instrumentalists grew rapidly and sometimes exceeded the supply. Working almost exclusively for small enterprise, musicians found lucrative, steady employment across the nation. In addition, they, unlike their employers, were effectively organized, which meant that employers were often incapable of resisting their collective demands. The problems musicians faced in this age were insignificant compared with those confronting other groups of skilled workers, many of whom were locked in bitter, even deadly disputes with employers over the course of industrial and technological development.

In the 1920s capitalist development, and more specifically the use of new technology under capitalist control, began reshaping the lives of musicians much as it had long been affecting workers in many mass-production industries. In music, however, mechanization did not speed up the pace of work or make work more routinized or monotonous, as it did in other industries; nor did it reduce the skill levels of employed musicians. Instead it eliminated major sources of employment with no regard to seniority and little regard to skill. More than twenty thousand musicians— approximately a quarter of the nation's professional instrumentalists and half of those who were fully employed—lost their jobs within a few years of the advent of sound movies; additional thousands lost jobs with the perfection of recording and radio technology and the industrial changes that such perfection permitted and even dictated. Unlike bookkeepers, copyists, and other workers displaced by innovations in office machinery, for example, musicians could not retrain themselves or transfer into new jobs that to one degree or another utilized their old skills. Displacement meant for them, more than for most workers, rapid devastation on a grand scale.[1]

Widespread job loss was only the direct effect of mechanization. As the labor shortage of the turn-of-the-century years became a surplus after 1928, the mood and outlook of working musicians changed. Mounting numbers of them found life more stressful, as they worried about their future and that of their craft. Some lost their sense of identity and self-control and retreated into social alienation. "We formerly owned and controlled our lives and profession by taking our instruments out of the cases and playing music," one of them said in 1942. "Now this ownership and control has been taken away from us . . . by these several mechanized systems."[2] Faced with these circumstances, many musicians, including most of those in certain types of employment, had no choice but to look for different lines of work. The fact that most of them had no other comparable work skills made their search the more difficult and poignant, especially during the Great Depression.

Although technological change eliminated musical jobs across the nation, it created new opportunities for musicians in media centers, where entrepreneurs took advantage of the business as well as the artistic "advances" the new technologies permitted. Thousands of instrumentalists, pushed by real economic pressures and pulled by potential employment opportunities, flocked to these centers to advance or even to save their careers as musicians. The fortunate few of them who secured jobs in film and radio studios found the work lucrative but the work environment constrained by novel patterns of hiring, definitions of skill, and divisions of labor. Through the American Federation of Musicians (AFM) they exercised meaningful control over this environment for a while, but the competition for jobs and the insecurities consequent upon that competition generated new anxieties for them as well as new problems for their union.

Musicians responded variously to these new circumstances. Some accepted the changes as inevitable consequences of progress; others struggled to adapt to or otherwise overcome the challenges those changes posed. Individual accommodation, however, was less important than collective reaction. In the late 1920s and 1930s, when innovation was most rapid and overwhelming, even the AFM was unsure how to respond to the problem of technological displacement. Under the leadership of Joseph N. Weber the union pursued largely unsuccessful tactics of accommodation, which gave way to a much more aggressive stance under the leadership of James C. Petrillo after 1940. Weber and Petrillo generally agreed that technological innovation was a fact of life in the music industry; but they differed radically in the ways they tried to influence conditions in the workplace itself.

The musicians' response to technological change cannot be gauged solely from their activities. On an unconscious as well as a conscious level, musicians sought to control the language of the discourse concerning their future. In trade papers, press statements, and union halls, they struggled to establish the legitimacy of their interests and perspectives, and thus of collective bargaining, union solidarity, and the strikes and bans that inconvenienced or otherwise adversely affected the public. In this battle for the high ground their leaders and spokesmen drew on longstanding American ideals as well as more immediate social fears. By the middle of the twenti-

Table 4 AFM Membership in Selected Locals, 1928–1942

City and Local	1928	1934	1936	1942	Growth during Period (%)
New York (802)	15,654	15,078	15,744	21,036	+34
Chicago (10)	7,146	7,084	7,026	9,685	+36
Los Angeles (47)	3,494	3,340	3,899	6,465	+85
Cleveland (4)	1,458	1,263	1,311	1,409	–3
San Francisco (6)	2,700	2,260	2,550	2,825	+5
Boston (9)	2,459	1,670	1,533	1,431	–42
Newark (16)	1,442	1,070	1,113	991	–31
Denver (20)	784	427	495	506	–35
San Antonio (23)	335	188	160	248	–26
Kansas City (34)	952	459	533	680	–29
Baltimore (40)	1,239	770	770	899	–27
Racine (42)	171	129	168	217	+27
Omaha (70)	581	339	331	345	–41
Memphis (71)	250	172	215	236	–6
Minneapolis (73)	1,148	569	757	1,195	+4
Philadelphia (77)	2,846	2,133	2,201	2,702	–5
Atlanta (148)	333	144	149	173	–48
Yonkers (402)	195	182	139	163	–16
Miami (655)	407	278	305	578	+42
Tampa (721)	190	127	102	138	–27

Source: Official Proceedings, 1928–42.

Note: This table sheds light on the decline and centralization of musical opportunities between 1928 and 1942. It indicates that the number of professional musicians in New York, Los Angeles, and Chicago increased significantly while the number in many other cities declined. In 1928 about 17 percent of total AFM membership lived in these three large media centers; by 1942 that figure had risen to 28 percent.

eth century this effort had largely failed. "The huge profits from mechanized music," as one musician noted at the time, "have gone not to the performing musicians, but to the middlemen controlling electronic transmission."[3]

That was not the entire truth. In the fight to mitigate the worst effects of technological change, musicians had in fact had a measure of success. After the disappearance of theater orchestras, their union protected their jobs in radio far longer than one might have expected; and the result of the effort amounted to much more than delaying disaster or stringing it out over a long period of time. In media centers like Los Angeles and New

York, the AFM's voice in labor relations was long-lasting, and the most long-lasting of its accomplishments was the royalty trust fund. In a period of political reaction and rising antiunionism, musicians forced upon management a revolutionary institutional arrangement for sharing the benefits of "modernization"; the Music Performance Trust Fund, which embodied that arrangement, has proved an enduring achievement. It is still an important source of income for AFM locals across the country. "Petrillo did some wonderful things for the music profession," one musician said with the fund in mind.[4]

All of those things met employer resistance. Through associations of their own, employers of musicians waged generally effective campaigns against the AFM. Petrillo may not have been exaggerating when he characterized the National Association of Broadcasters' public relations blitz during the first recording ban as "the strongest ever used to arouse the public against union officials."[5] To win public support for their position, employers manipulated symbols and slogans of their own. Through the media they, more effectively than the union, appealed to popular notions of progress, patriotism, and democracy, and in doing so they convinced the public that the musicians' demands were excessive, even irrational. What was truly irrational, however, was to expect musicians to stand passively by as their jobs disappeared.

Employers in mass-entertainment industries saw their situation much as their counterparts in other industries had seen theirs when they had earlier confronted technological transformation. Theirs, they believed, were the usual prerogatives of ownership and management, including control over production. To them new technology meant increased efficiency and thus progress. It meant better control of production processes and thus higher quality and more uniformity in the product they produced. More important, it lowered labor costs, sometimes fabulously so. Musicians, employers believed, had to adapt, as artisan shoemakers and other craftsmen had earlier had to adapt, to the social change dictated by technological innovation. Such change was the cost of progress. "The cold facts are that motion picture production had to change to meet the public demand for sound movies," film producers explained, "or else the industry would have gone backward with unpredictable consequences to both labor and management."[6]

These attitudes reflected more than greed and insensitivity. Technological innovation in fact presented businessmen with challenges of their own, and those without adequate financial resources or managerial skills them-

Table 5 Employment of Staff Musicians in Radio, 1946–1957

Year[1]	Staff Musicians[2]	Earnings	Stations
1946	2,433	$12,056,653	292
1947	2,230	11,110,457	292
1948	2,193	10,789,841	300
1949	2,073	10,277,052	301
1950	1,929	10,073,272	305
1951	1,739	11,326,028	248
1953	1,610	10,247,406	213
1954	1,267	9,374,776	171
1955	1,164	9,126,656	151
1956	1,011	6,084,972	120
1957	576	3,483,946	98

Source: *Official Proceedings*, 1946–57.

[1] Figures for 1952 were not available.

[2] These musicians worked fifty to fifty-two weeks a year. Several hundred more instrumentalists each year worked under contracts guaranteeing employment for thirty to fifty hours a week.

selves fell by the wayside. In the highly competitive entertainment business, innovation created tensions between firms and encouraged predatory behavior as well. Those things encouraged competing businesses to displace workers in order not just to prosper but to survive. The Great Depression added to these pressures.[7]

Yet the triumph of recorded music was possible, in the final analysis, only because consumers came to prefer, and then to demand, the superior product it made available to them at an almost nominal cost. Like moviegoers who lined up for talkies, radio audiences tuned in to the music of big-city bands, regardless of whether it was transcribed or live. Consumers were indifferent to the concerns of musicians, and even hostile to them when they threatened the supply of recorded and broadcast music. The issue, then, was largely decided by marketplace forces.

Nonetheless, political interference with those forces was an important influence on the final outcome. In the postwar years a Congress intent on reducing labor-management conflict outlawed a number of trade union practices that musicians had relied upon to protect themselves in the marketplace. "When they passed those laws," union representative Phil Fischer said of the Lea and Taft-Hartley acts, "we lost our club. We could no longer demand employment."[8] The course of events testified to the truth

Stage to Studio

Table 6 Single-Engagement Commercial Broadcasting Employment, 1946–1957

Year[1]	Number of Sponsored Programs[2]	Earnings of Leaders and Sidemen[3]
1946	337	$8,213,787
1947	370	7,695,699
1948	323	7,177,246
1949	248	5,372,483
1950	194	4,733,574
1951	181	4,651,152
1953	179	2,605,518
1954	158	2,089,758
1955	250	2,684,690
1956	101	849,718
1957	63	452,659

Source: Official Proceedings, 1946–57.

[1] Figures for 1952 were not available.

[2] The vast majority of these programs originated in New York and Los Angeles; others originated in Chicago, Boston, Nashville, San Francisco, and a few smaller cities.

[3] The total number of musicians earning wages in commercial radio is unclear because instrumentalists often worked for more than one sponsored program.

of those words. In the 1950s, when competition from television and tape recording presented new incentives for displacing musicians, job opportunities in radio all but disappeared. By the end of the decade radio employed barely 350 full-time musicians and tendered less than $1 million from single-engagement employment.[9]

Government-business relations suggested that lawmakers and industrialists had more affinity for each other than either had for musicians. This affinity stemmed partly from the financial investments some legislators had in media industries, but it also stemmed from common ideological assumptions about economics, technology, and the public interest. All lawmakers no doubt sympathized at some level with the problems of technologically displaced workers, but they also accepted the legitimacy of what they ultimately regarded as the rights of property. It was probably the latter consideration that best explains the actions of legislators.

Musicians, on the other hand, believed that lawmakers' actions did violence to their rights as citizens and their interests as workers. The "object of a true democratic form of government," Weber said back in 1938, is to "protect against unemployment and its resulting misery," especially to protect victims of technological displacement. "Government cannot shirk its duty to care for the worker who is unemployed against his will." Petrillo shared that view. In 1942 he told senators that if employers would not "share the profits" of technological innovation, he as leader of a union of workers displaced by such innovation had a right to use the collective bargaining process to encourage them to do so, and a right to do that without government interference. Insofar as government had a legitimate interest in this process, he added, it was to help musicians save their livelihood and their contribution to American culture.[10]

FOR WORKING MUSICIANS as a collectivity, the impact of technological change was devastating. In the new world that emerged from that change, only the most mobile, talented, and well-connected musicians prospered. Indeed, that group prospered as few musicians had ever prospered. This experience speaks to familiar patterns in the history of American workers. Over and over American capitalism has achieved economic growth and increased material consumption by devaluing the skills and bargaining power of specific groups of workers while advancing the well-being of other, often much smaller, groups of workers.

This pattern raises troubling questions for us today. Technological change is one of the overriding realities of our times; its economic and social potentialities, good and bad, are part of our socioeconomic reality. Unfortunately, corporate America has convinced the general public that technological change is inherently emancipatory, not enthralling. "Automation is the magical key to the creation," the National Association of Manufacturers pronounced in 1955. "[With] the smooth, effortless workings of automation [and] the cooperation of Americans in all walks of life, our standard of living will skyrocket, prices drop, markets expand and the tempo of prosperity accelerate. More workers will be needed in the fields of recreation and amusements [because] the work week will shorten, the hours of leisure lengthen."[11] But the fate of musicians in the wake of the technological changes that displaced them on a massive scale suggests an alternative scenario, one in which innovation ravages the lives as well as the livelihoods of masses of people, even highly skilled members of a profession.

As long as industrial technology is designed and used primarily for increasing productivity at the lowest possible price, it will remain a sharp, double-edged sword. Though it will generate demands for new skills and talents and increase the status—and pay—of workers who fill those demands (even if it slots those workers into narrow, task-oriented jobs), it will leave others, in the words of one historian, "to vegetate in the backwaters of

Stage to Studio

the stream of progress."[12] Those prospects will continue until society comes to view technological change as a social problem as well as a matter of labor productivity. This is not to say that businesses must retain obsolete jobs, or that government should stop supporting research and development. It is to say, rather, that our well-being as a society depends in part at least on spreading the effects of mechanization more evenly among employers, investors, consumers, and workers. We must find better ways to cope with the negative implications of technological change. Change inevitably means that some workers lose their jobs, but it need not mean that they lose their pride and power as well, and it certainly should not reduce their stake in society. By insisting that those who benefit from new technology help those who do not, perhaps we can find new ways to ease the transition from one workplace to another, and even increase our leisure time.

Appendix: AFM Membership, 1896–1956

THE FOLLOWING GRAPHS are based on *Official Proceedings,* 1896–1956; various issues of *International Musician* and *Overture;* Everett Lee Refior, "The American Federation of Musicians: Organization, Policies, and Practices" (master's thesis, University of Chicago, 1955), 55; and Robert D. Leiter, *The Musicians and Petrillo* (New York: Book-man Associates, 1953), 80.

Total membership figures for 1901–3, 1911–12, 1914, and 1919–28 are contradictory or not available. For the years 1901–3, 1911–12, and 1914, membership figures in the graph are linearly interpolated from nearby known data. For 1919–28, figures were obtained by linear extrapolation from membership of ten representative locals.

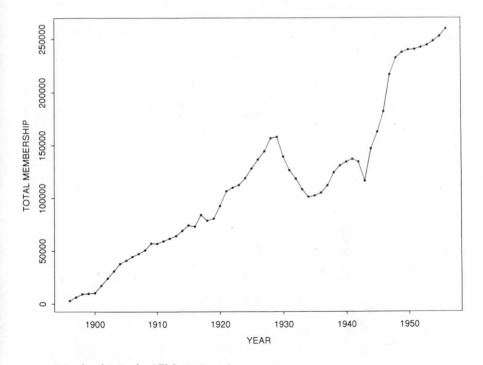

Membership in the AFM, 1896–1956

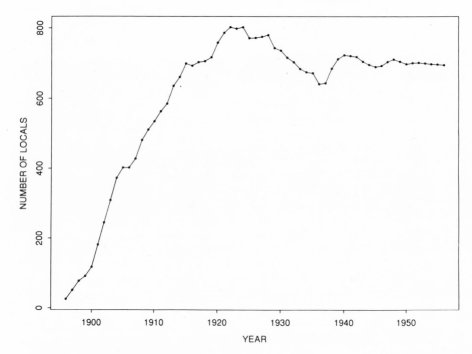

Number of Locals in the AFM, 1896–1956

Appendix

Notes

Abbreviations

AFM Microfilm, OHS   American Federation of Musicians records, Microfilm
                       Department, Archives–Library Division, Ohio Historical
                       Society, Columbus

Bagley Collection   Charles Leland Bagley Collection, Special Collections Li-
                       brary, University of Southern California, Los Angeles

Official Proceedings   *Official Proceedings* of the annual conventions of the
                       American Federation of Musicians, in Bagley Collection;
                       cited by year

Papers of Local 47   Papers of the Los Angeles Musicians' Union, Local 47 of
                       the AFM, President's Office File, at the local's headquar-
                       ters in Los Angeles

Introduction

1. "Report of Records and Transcriptions Committee," poster in Bagley Collection.

2. Among recent studies of the relationship between work and technological change, see Harry Braverman, *Labor and Monopoly Capital: The Degradation of Work in the Twentieth Century* (New York: Monthly Review Press, 1974), which emphasizes that new technology separated the "conception" of work from its actual "execution." Richard Edwards, *Contested Terrain: The Transformation of the Workplace in the Twentieth Century* (New York: Basic Books, 1979), explains that the rise of impersonal bureaucracies increased management's control over labor. Both Braverman and Edwards raise the question of whether management deployed new technology to increase production or to gain more control over the workforce. See also Michael Burawoy, *The Politics of Production* (New York: Verso, 1985); and David Noble, *Forces of Production: A Social History of Industrial Automation* (New York: Oxford University Press, 1984).

3. Two of the most frequently cited works focusing on labor's response to technological change are David Brody, *Steelworkers in America: The Non-Union Era* (Cambridge: Harvard University Press, 1960); and David Montgomery, *Workers' Control in America* (New York: Cambridge University Press, 1979).

4. For a general overview of the literature on government-business relations, see Thomas K. McCraw, "Regulation in America: A Review Article," *Business History Review* 49 (Summer 1975): 159–83; Louis Galambos, "Technology, Political Economy, and

Professionalization: Central Themes of the Organizational Synthesis," *Business History Review* 57 (Winter 1983): 471–93; and Alan Dawley, "Workers, Capital, and the State in the Twentieth Century," in *Perspectives on American Labor History: The Problems of Synthesis,* ed. J. Carroll Moody and Alice Kessler-Harris (DeKalb: Northern Illinois University Press, 1990), 152–200.

5. Robert D. Leiter, *The Musicians and Petrillo* (New York: Bookman Associates, 1953), which examines the policies of AFM leadership, has been the best work on the history of the union for nearly half a century. George Seltzer, *Music Matters: The Performer and the American Federation of Musicians* (Metuchen, N.J.: Scarecrow, 1989), offers a more comprehensive study of the union. Various theses and dissertations cited in this study complement these works.

6. See Sandy R. Mazzola, "When Music Is Labor: Chicago Bands and Orchestras and the Origins of the Chicago Federation of Musicians, 1880–1902" (Ph.D. diss., Northern Illinois University, 1985).

7. Robert Heilbroner, *The Economic Transformation of America: 1600 to the Present,* 3d ed. (New York: Harcourt Brace and Company, 1994), 12.

8. For an overview of "postmaterialist" interpretations of labor history, see Lenard R. Berlanstein, ed., *Rethinking Labor History: Essays on Discourse and Class Analysis* (Urbana: University of Illinois Press, 1993). A related work that focuses on the ideology of business is Howell John Harris, *The Right to Manage: Industrial Relations Policies of American Business in the 1940s* (Madison: University of Wisconsin Press, 1982).

Chapter One: Working Scales in Industrial America

1. Margaret Hindle Hazen and Robert M. Hazen, *The Music Men: An Illustrated History of Brass Bands in America, 1800–1920* (Washington, D.C.: Smithsonian Institution Press, 1987), 41–69.

2. Craig H. Roell, *The Piano in America, 1890–1940* (Chapel Hill: University of North Carolina Press, 1989), 10–12.

3. Hazen and Hazen, *Music Men,* 30, 37.

4. Sandy R. Mazzola, "When Music Is Labor: Chicago Bands and Orchestras and the Origins of the Chicago Federation of Musicians, 1880–1902" (Ph.D. diss., Northern Illinois University, 1985), 193, 208–10.

5. Ibid., 18. The primary source of data for this analysis was occupational statistics of the U.S. Census for 1870, 1880, 1890, and 1900. See *Compendium of the Ninth Census* (Washington, D.C., 1872), 604–7, table 65; *Compendium of the Tenth Census* (Washington, D.C., 1883), 1368–69, table 103; *Compendium of the Eleventh Census* (Washington, D.C., 1897), 396, table 74; and *Occupations at the Twelfth Census* (Washington, D.C., 1904), l, table 4. Only the 1870 census distinguished between "musicians" and "music teachers." In that year there were 6,519 musicians (only 173 of whom were women) and 9,491 music teachers (5,580 women). Subsequent census reports combined these two categories.

6. Vern L. Countryman, "The Organized Musicians" (part 1), *University of Chicago Law Review* 16 (Autumn 1948): 57; Robert D. Leiter, *The Musicians and Petrillo* (New York: Bookman Associates, 1953), 13.

7. Countryman, "Organized Musicians," 57; Leiter, *Musicians and Petrillo,* 12.

8. John R. Commons, "The Musicians of St. Louis and New York," in *Labor and Administration* (New York: Macmillan, 1913), 300, reprinted from *Quarterly Journal of Economics* 20 (May 1906), 419–42; Ernest J. Englander, "The Inside Contract System of Production and Organization: A Neglected Aspect of the History of the Firm," *Labor History* 28 (Fall 1987): 429–46.

9. *Constitution and By-laws of the Washington Musical Protection Union,* 1865, in Bagley Collection; Body Meetings from January 1904 to December 1913, 495, Papers of Boston Musicians' Association, Local 9 of the American Federation of Musicians, Boston. Also Price List of the American Federation of Musicians, Local 103, MIC 155, vol. 1 (1891), 19–20, AFM Microfilm, OHS.

10. *Constitution and By-laws of the Washington Musical Protection Union,* 1865, in Bagley Collection; Price List of the American Federation of Musicians, Local 103, MIC 155, vol. 1 (1891), 23, AFM Microfilm, OHS. Leiter states that musicians were "barely able to eke out a living." Leiter, *Musicians and Petrillo,* 11.

11. William P. Steeper, "Civil Rights in the American Federation of Musicians," *International Musician,* December 1954, special section, 22.

12. The 1865 constitution of the Washington Musical Protective Union stated that members must "refuse to perform in any orchestra or band in which any person or persons are employed who are not members of this Union." *Constitution and Rules of Order of the Washington Musical Protective Union.* (Washington, D.C.: Gibson Brothers, 1865), 6, in Bagley Collection. The 1906 constitution of the Los Angeles musicians' union Local 47 forbade members to supply nonunion musicians with information regarding the union price list and directory. A fine of up to $100 could be levied against violators. In 1924 the fine was up to $200. Members had to obtain the union's permission even to teach or conduct nonunion musicians. *Constitution and By-laws of the MMPA of Los Angeles,* 1906, 20, and *Constitution and By-laws,* 1924, in Bagley Collection.

13. Countryman, "Organized Musicians," 76; Robert Lee Humes, "Labor Relations and the American Federation of Musicians: Six Locals in Pennsylvania" (master's thesis, Pennsylvania State University, 1965), 67. Humes indicates that some unions required applicants to obtain recommendations from two union members, usually bandleaders. Local amateur bands, variously organized by fire or police departments, American Legion posts or other patriotic groups, and public or private schools, were sources of competition for union musicians. Employers often preferred amateur bands because they worked free or for only nominal sums. Union objections to the use of such bands often aroused the ire of the community, but union officers always objected to amateur performances that took jobs away from union musicians. Unions thus worked to limit performances by amateur groups to occasions that were not competitive or commercially oriented, such as church functions and patriotic occasions. See *Constitution and By-laws of the MMPA of Los Angeles,* 1937, in Bagley Collection; and letters from union officials to local musical groups, 1934–41, Papers of Local 47.

14. "The Musical Mutual Protective Association, 1887–1912," clipping in Bagley Collection.

15. Sue-Ellen Hershman, "What's in a Number? The History and Merger of Local 535," *Interlude,* January–February 1993, 4; Donald Spivey, *Union and the Black Musi-*

cian: The Narrative of William Everett Samuels and Chicago Local 208 (New York: University Press of America, 1984), 9–10, 53; Lewis A. Erenberg, *Steppin' Out: New York Nightlife and the Transformation of American Culture, 1890–1930* (Chicago: University of Chicago Press, 1981), 152.

16. Pianist Samuel Johnson, violinist George Bridgetower, and guitarist Justin Holland, all African Americans, were among the nation's best-known concert musicians in the late nineteenth century, when several black women also gained fame as vocalists. Marie Selika and Elizabeth Taylor Greenfield, for example, distinguished themselves as concert sopranos, the latter once singing before Queen Victoria; Spivey, *Union and the Black Musician*, 5–7. See also William Barlow, *Looking up at Down: The Emergence of Blues Culture* (Philadelphia: Temple University Press, 1989), 119; and Susan Curtis, *Dancing to a Black Man's Tune: A Life of Scott Joplin* (Columbia: University of Missouri Press, 1994).

17. Barlow, *Looking up at Down*, 54–55, 116–17.

18. Ibid., 61–63. In the early twentieth century Thomas apparently recorded more than twenty of his songs for some of the nation's first recording companies.

19. Price List of the American Federation of Musicians, Local 103, MIC 155, vol. 4 (1909), 23, AFM Microfilm, OHS; *Monthly Labor Review*, November 1931, 1014–16. See also "An A.F.L. View of Women Workers, 1897," in Eileen Boris and Nelson Lichtenstein, eds., *Major Problems in the History of American Workers: Documents and Essays* (Lexington, Mass.: D. C. Heath, 1991), 232–33.

20. *Eleventh Census* (1890), 396, table 74. A few women served as secretary-treasurers of local unions. See George Seltzer, *Music Matters: The Performer and the American Federation of Musicians* (Metuchen, N.J.: Scarecrow, 1989), 213–14; *Overture*, May 1950, 10; and April 1950, 38. See also Roell, *Piano in America*, 13–16. The fact that women could and did rely on one another to learn to play musical instruments no doubt encouraged women to study music.

21. Roell, *Piano in America*, 14–17. See also Barbara Welter, "The Cult of True Womanhood, 1820–1860," *American Quarterly* 18 (1966): 151–74.

22. *Ninth Census* (1870), 604–7, table 65; *Tenth Census* (1880), 1368–69, table 103. Census reports of later years did not indicate musicians' nativity. Obituaries in the Chicago musicians' trade paper, *Intermezzo* (June 1940, 10, and October 1940, 11), offer insight into the ethnic diversity of musicians in the late nineteenth and early twentieth centuries. Of eight musicians mentioned, three were American-born; two were born in Czechoslovakia, one in Norway, and one in Poland. The birthplace of another was unknown. Most had been longtime members of Chicago Local 10. All issues of *Intermezzo* cited in this study are located in AFM Local 10, Chicago.

23. "The Musical Mutual Protective Association, 1887–1912," clipping in Bagley Collection.

24. Ibid.

25. Abram Loft, "Musicians, Guild, and Union: A Consideration of the Evolution of Protective Organization among Musicians" (Ph.D. diss., Columbia University, 1950), 289. Loft reports that the MNPA treasury handled $504.80 from 1872 to 1875. The MNPA never had more than seventeen affiliates. *Official Proceedings*, 1947, 279.

26. *American Musician*, May 1, 1898, 2. Also *Official Proceedings*, 1947, 279. Leiter notes that the NLM had fifteen locals in 1887; in 1896 it had seventy-nine locals repre-

senting nine thousand musicians. Leiter, *Musicians and Petrillo*, 14. Loft's dissertation gives a more detailed account of the history of the NLM. See Loft, "Musicians," 357–87. See also Seltzer, *Music Matters*, 5–6.

27. Countryman, "Organized Musicians," 58.

28. Owen Miller, "An History of the American Federation of Musicians," 2–3, clipping in Bagley Collection. Also *Official Proceedings*, 1947, 247.

29. For a look at the "duality" of musical culture, see Michael Broyles, *"Music of the Highest Class": Elitism and Populism in Antebellum Boston* (New Haven: Yale University Press, 1992); and Lawrence W. Levine, *Highbrow/Lowbrow: The Emergence of Cultural Hierarchy in America* (Cambridge: Harvard University Press, 1988), 85–168.

30. Owen Miller to Samuel Gompers, March 19, 1897, American Federation of Labor Records, The Samuel Gompers Era, microfilm edition, reel 141, Special Collections Library, University of California, Los Angeles. "Bremer and Ruhe," Miller told Gompers, "are not representing the mass of musicians by any means."

31. Loft, "Musicians," 373. Miller insisted that locals "must sacrifice some autonomy to strengthen the national body." NLM leaders often wore Prince Albert coats and silk hats to annual conventions; hence their nicknames. *International Musician*, May 1940, 1.

32. Loft, "Musicians," 377; *American Musician*, May 1, 1898, 3–4. Gompers actually began admitting local unions of musicians to the AFL in 1891.

33. Steeper, "Civil Rights in the American Federation of Musicians," 6. Specific duties of the president included presiding at all meetings of the executive board, signing official documents, appointing all committees, and reporting on his activities in an annual report. The ability of the president to annul provisions of the constitution has been cited as evidence that the union was a dictatorship. In fact, this power could be used only under special circumstances; it was used only a few times and for such purposes as setting aside dues payments for members in the armed forces, or for changing the dates of conventions. Locals repeatedly insisted on keeping this provision intact.

34. *American Musician*, May 1, 1898, 3–4, and Miller, "History of the AFM," clipping in Bagley Collection. Also John Scott Kubach, "Unemployment and the American Federation of Musicians: A Case Study of the Economic Ramifications of Technological Innovations and Concomitant Governmental Policies Relative to the Instrumental Employment Opportunities of the Organized Professional Musicians" (master's thesis, Ohio University, 1957), 35–36. Delegates of the first AFM convention agreed that in jurisdictions where two locals existed, the locals must amalgamate within three months, and AFL officials would grant a charter to the local making the most legitimate effort to consolidate. As the AFM's first president, Miller drew up plans for the union's second national convention, held in Kansas City in May 1897.

35. The union gave no local more than ten votes at annual conventions. Also, provisions of the constitution could be amended or repealed by majority vote at the annual convention. (In 1906 the ten-vote maximum rule was amended to give larger locals more power in the enactment of new laws.) Furthermore, the AFM gave more autonomy to local unions. Not only did each local set its own wage scales and rules regarding working conditions, but it also maintained the right to discipline or expel members so long as a formal hearing was held. Members had the right to appeal the local's decision to the national executive board. Countryman, "Organized Musicians," 59.

36. Ibid. See also Loft, "Musicians," 382. The Jackson County District Court issued the injunction.

37. *American Musician*, May 1, 1898, 2–4. The league distributed its remaining funds at the 1904 convention. Relations between the federation and its New York local remained testy for nearly two decades. In 1921, when New York refused to accept transfer cards from traveling musicians, the federation suspended its New York branch and chartered a new local in the city, Local 802. The new union quickly absorbed the membership of its predecessor, Local 310, and assumed control over labor relations in the New York music sector. For a closer look at the evolution of New York locals, see Leiter, *Musicians and Petrillo*, 28–32.

38. Miller, "History of the AFM," 4, clipping in Bagley Collection; Countryman, "Organized Musicians," 59. Also Humes, "Labor Relations and the American Federation of Musicians," 4; and Leiter, *Musicians and Petrillo*, 80.

39. Leiter concluded that in 1905 the AFM "had developed a position of more complete control over its business . . . than that occupied by any other union in the American Federation of Labor." Leiter, *Musicians and Petrillo*, 24–25.

40. Kubach, "Unemployment and the American Federation of Musicians," 35. Miller's salary was probably supplemented by wages he earned as president of AFM Local 2 in St. Louis. He apparently did not begin his duties as editor of *International Musician* until 1904.

41. *International Musician*, December 1950, 7. The date of Weber's birth is unclear, since one source states that he was born in 1865. It is also unclear how old he was when he came to America; two sources say fourteen, another says nine. In 1929 Weber was elected vice president of the AFL.

42. Ibid., August 1940, 1; December 1950, 7; Seltzer, *Music Matters*, 12–14; Kubach, "Unemployment and the American Federation of Musicians," 36. Seltzer provides sketches of other individuals who played prominent roles in the formation of the AFM. His picture suggests a strong German American influence. The first vice president was German-born George Nachman, who worked in the Odeon Theater in Baltimore for twenty-six years. The second vice president was Christian Ahbe, violinist and cornet player, also German-born. Ahbe had been a member of the Voss' First Regiment Band in Newark, New Jersey, for nearly twenty-five years. The first secretary of the AFM was Swiss-born Jacob J. Schmalz, also a performing musician. In the 1870s Schmalz apparently moved to Cincinnati, where he played an active role in the formation of the AFM. He had been one of the founders of the NLM.

43. *International Musician*, August 1940, 1.

44. *Constitution and By-laws of the Musical Protective Association*, 1872, 23, in Bagley Collection. Civilian protests of military performances dated back to 1824, when the War Department began granting extra leaves of absences for enlisted musicians. The increased opportunities for work (and income), the department hoped, would induce more musicians to join and remain in the service.

45. *Official Proceedings*, 1932, 36–37. The Army and Navy Appropriation Acts increased the pay of military bands. See also Countryman, "Organized Musicians," 73; Leiter, *Musicians and Petrillo*, 38–40; and Commons, "Musicians of St. Louis and New York," 316–17. Commons explains that the presence of one military band in St. Louis "at all times compelled [civilian bands] to adjust their prices to what managers said they could get the Army for."

46. Countryman, "Organized Musicians," 74. Circuses and fairs were especially likely to hire foreign musicians. When American cities held great fairs at the turn of the century, the AFM sent circulars to musical circles in Europe warning of unscrupulous agents who contracted musicians without mentioning America's high cost of living. Loft, "Musicians," 302. The MMPU adopted a contract labor policy of its own. Members were prohibited from performing with any foreign musician for six months after the foreigner's arrival. *New York Herald,* July 16, 1891, clipping in Bagley Collection.

47. *New York Daily Journal,* undated clipping in Bagley Collection; *Official Proceedings,* 1932, 33–35. In 1932 AFM president Joseph Weber told Congress that the law was hypocritical and did not reflect the intention of the original lawmakers. "All an emigrant has to do," Weber said, "is put a musical instrument under his arm, claim to have a musical engagement here, and remain here, even though he may desert his music the next day." Such interpretation of the law, he explained, led to greater competition for American laborers in other trades. In demanding an amendment to the 1885 law, Weber said, "Music is an art, but it does not follow that every musician is an artist." Ninety-five percent of immigrant musicians, he suggested, were not. On March 17, 1932, President Herbert Hoover signed House Bill Number 8335, which amended the contract labor law.

48. Loft, "Musicians," 361–62; *Official Proceedings of the Twelfth Annual Convention of the NLM,* 14, in Bagley Collection. The NLM had similarly accommodated what amounted to national membership by allowing members to transfer from one affiliate to another for a small fee, which practice was common in all craft unions at the time.

49. *Constitution and By-laws of the MMPA of Local 47, Los Angeles,* 1911, in Bagley Collection. By 1920 traveling bands were required to charge employers a 30 percent surcharge in excess of local union wage rates and to charge for transportation costs "by railroad or otherwise." The Thirty Percent Law, which applied only to engagements of a week or longer, was superseded in 1934 by the Ten Percent Law, which demanded an extra 10 percent collection over standard wage rates for every engagement, even one lasting a single day. Surcharges were split among the members who played the engagement, the local in whose jurisdiction the engagement was played, and the national headquarters. "Notice to all Locals," from Harry Brenton, treasurer of AFM, 1934, and "Notice to all Locals," from AFM president Weber, February 26, 1935, Papers of Local 47.

50. Spivey, *Union and the Black Musician,* 9–10, 52–54; Hershman, "What's in a Number?" 5; *International Musician,* July 1944, 1.

51. Hershman, "What's in a Number?" 4–5. The first president of Local 535 was Bill Smith; the first secretary-treasurer was Harry Hicks. Hershman notes that Local 535 evolved from meetings in a Boston music store and booking office.

Chapter Two: Boom and Bust in Early Movie Theaters

1. *The Motion Picture* 2, no. 5, p. 3, clipping in Bagley Collection. See also *New York Times,* April 24, 1895.

2. *Official Proceedings,* 1929, 43–44. In 1913 sidemen traveling with vaudeville shows earned approximately $50 a week, and leaders $75. Body Meetings from January 1904 to December 1913, 495, Papers of Boston Musicians' Association, Local 9 of the Ameri-

can Federation of Musicians, Boston (hereafter cited as Papers of Local 9). Also Edward Renton, *The Vaudeville Theatre: Building Operation Management* (New York: Gotham, 1918), 296–97; and Douglas Gilbert, *American Vaudeville: Its Life and Times* (New York: Whittlesey House, 1940), 32. Quotation from Q. David Bowers, *Nickelodeon Theatres and Their Music* (Vestal, N.Y.: Vestal, 1986), 127, 129.

3. *Billboard,* January 3, 1942. See also Douglas Gomery, *Shared Pleasures: A History of Movie Presentation in the United States* (Madison: University of Wisconsin Press, 1992), 18–33; Russel Merritt, "Nickelodeon Theaters, 1905–1914: Building an Audience for the Movies," in *The American Film Industry,* ed. Tino Balio (Madison: University of Wisconsin Press, 1976), 59–82; Craig H. Roell, *The Piano in America, 1890–1940* (Chapel Hill: University of North Carolina Press, 1989), 49–52; and Bowers, *Nickelodeon Theatres and Their Music,* 129–31, 171.

4. Quotation from Roell, *Piano in America,* 52. On photoplayers, see Bowers, *Nickelodeon Theatres and Their Music,* 131–70. Operators could even replace music rolls while the instruments played, although this was not easy.

5. Roell, *Piano in America,* 52; Bowers, *Nickelodeon Theatres and Their Music,* 131–70. Photoplayer sales were strongest from 1913 to 1923, before the introduction of large theater organs.

6. Gomery, *Shared Pleasures,* 32–56; Bowers, *Nickelodeon Theatres and Their Music,* 115; Lizabeth Cohen, *Making a New Deal: Industrial Workers in Chicago, 1919–1939* (Cambridge: Cambridge University Press, 1990), 124–30.

7. *Marque* 23 (Third Quarter 1991): 27–31. The growing size of orchestras necessitated increased support staff. New York's palatial Paramount increased its "musical faculty" to include three orchestrators, three librarians, and a voice coach. Ben M. Hall, *The Best Remaining Seats: The Story of the Golden Age of the Movie Palace* (New York: Clarkson N. Potter, 1961), 175.

8. *Official Proceedings,* 1928, 44; *Exhibitor's Herald,* October 20, 1928, 52. On air conditioning, see Gomery, *Shared Pleasures,* 53–54.

9. *The Motion Picture* 2, no. 5, p. 3, clipping in Bagley Collection; *Official Proceedings,* 1930, 28; 1955, 123; *Film Daily,* October 28, 1929, 1; *International Musician,* January 1954, 34; *Intermezzo,* April 1928, 9; *Monthly Labor Review* 33 (August 1931): 3. See also Robert D. Leiter, *The Musicians and Petrillo* (New York: Bookman Associates, 1953), 56–57, 80; and John Scott Kubach, "Unemployment and the American Federation of Musicians: A Case Study of the Economic Ramifications of Technological Innovations and Concomitant Governmental Policies Relative to the Instrumental Employment Opportunities of the Organized Professional Musicians" (master's thesis, Ohio University, 1957), 207, 214, 219. After 1870 the U.S. Census did not distinguish between "musicians" and "music teachers." The 1930 figure for "musicians and music teachers" was 165,128 (85,517 men and 79,611 women). *Fifteenth Census of the United States: 1930* (Washington, D.C., 1933), 5:20.

10. *Monthly Labor Review,* 33 (November 1931): 1006–7; *Film Daily,* October 22, 1929, 1; *Overture,* November 1926, 14; *Official Proceedings,* 1929, 14; *International Musician,* January 1954, 34. In 1928 Weber estimated that ten thousand musicians held full-time jobs in hotels, cafés, and restaurants, and that another fifteen hundred worked in traveling bands. *Intermezzo,* April 1928, 9. In Columbus, theater musicians in 1891 earned $16 a week (with or without Sunday performances), and their leaders

made $25. In 1920, theater musicians in Columbus could earn $45.50 a week, and their leaders $68.25. Price Lists of the American Federation of Musicians, Local 103, MIC 155, vol. 1 (1891), 24, and MIC 155, vol. 11 (1921), 5, AFM Microfilm, OHS.

11. *Official Proceedings*, 1918, 228; 1929, 46. President Weber reported 158,000 AFM members in 1929. Also Vern L. Countryman, "The Organized Musicians" (part 1), *University of Chicago Law Review* 16 (Autumn 1948): 57; Leiter, *Musicians and Petrillo*, 16, 80.

12. *Overture*, September 15, 1926, 3. *Overture*, the trade paper of AFM Local 47, was published twice a month until 1927, when it changed to a one-a-month format. See also MIC 155, vol. 11 (1921), 3, and vol. 6 (1914), 23, AFM Microfilm, OHS.

13. Body Meetings from January 1904 to December 1913, 465–67, Papers of Local 9; *Overture*, February 1936, 1.

14. *Overture*, September 15, 1926, 1, 6; September 1927, 2. By 1927 *Overture* had changed from a bimonthly to a monthly trade paper.

15. Rudy Behlmer, "Tumult, Battle, and Blaze: Looking Back on the 1920s—and Since—with Gaylord Carter, the Dean of Theater Organists," in *Film Music I*, ed. Clifford McCarty (New York: Garland, 1989), 19–24. Substitutes, Carter said, "were usually very good musicians."

16. *Overture*, January 1927, 16–17; March 1929, 15; *Exhibitor's Herald*, January 22, 1927, 35–36.

17. Roy M. Prendergast, *Film Music: A Neglected Art*, 2d ed. (New York: W. W. Norton and Company, 1992), 9–10; Tony Thomas, *Music for the Movies* (New York: A. S. Barnes and Company, 1973), 35–39; Behlmer, "Tumult, Battle, and Blaze," 28–29.

18. Hall, *Best Remaining Seats*, 185–97. The Marr and Colton Company manufactured an organ that activated preset combinations of different instruments to help organists generate specific moods. Stop keys were labeled "Hatred," "Sorrow," "Jealousy," "Childhood," "Happiness," "Rural," Oriental," and "Gruesome." It offered three types of "Love"—"Romantic," "Passion," and "Mother." Interview with Robert Alder, Honolulu, August 6, 1994.

19. Harry M. Geduld, *The Birth of the Talkies: From Edison to Jolson* (Bloomington: Indiana University Press, 1975), 37; Hall, *Best Remaining Seats*, 183; Behlmer, "Tumult, Battle, and Blaze," 47–48; Alder interview, August 6, 1994. Wurlitzer sold approximately two thousand of these versatile instruments from 1915 to 1935. Bowers, *Nickelodeon Theatres and Their Music*, 170–89.

20. Since keys occasionally failed and wires sometimes loosened, organists often had to repair their instruments on the spot. See Hall, *Best Remaining Seats*, 183, 198. Also *Musician*, November 1919, 8; *Overture*, August 1928, 21–22; September 1930, 10; June 1930, 22; and Dennis James, "Performing with Silent Films," in McCarty, *Film Music I*, 65–66. Some of the more famous theater organists included Jesse Crawford, Albert Hay Malotte, Henry Murtagh, and Oliver G. Wallace. Address by Farny R. Wurlitzer to American Association of Theatre Organ Enthusiasts, July 6, 1964, in Bowers, *Nickelodeon Theatres and Their Music*, 200–203.

21. Interview with Helen Lee, Los Angeles, June 21, 1989.

22. Ibid.

23. Ibid.

24. *Overture*, August 15, 1926, 11.

25. Ibid., July 1, 1926, 8; July 15, 1926, 8. Union bylaws typically demanded that if management required musicians to wear special costumes, management provided the costumes and paid each musician an additional fee of $2 a week to wear them. "Blacking up" to appear as "colored" musicians required an additional fee of $2 per performance. MIC 155, vol. 11 (1921), 1–3, and vol. 3 (1903), 23, AFM Microfilm, OHS.

26. "The Theater Is Perishing," *Theater Magazine,* March 1929, 19; *Overture,* April 1928, 2.

27. *Overture,* September 1930, 2; July 1, 1926, 18. Wages in dance halls, restaurants, and hotels also improved during this period. Leiter, *Musicians and Petrillo,* 80.

28. Warners advertised its name only as "Warner Bros." Joel Swenson, "The Entrepreneur's Role in Introducing the Sound Motion Picture," *Political Science Quarterly* 63 (September 1948): 411. See also McCarty, *Film Music I,* 74–77.

29. Warners played a key role in the introduction of sound by gambling $5 million on sound technology in the face of an indifferent and hostile silent-film industry. See Douglas Gomery, *The Hollywood Studio System* (New York: St. Martin's, 1986), 101–23; Swenson, "Entrepreneur's Role," 426; and Robert Sklar, *Movie-Made America: A Cultural History of the Movies* (New York: Chappell and Company, 1978), 152.

30. Douglas Gomery, "The Coming of Sound to the American Cinema: Transformation of an Industry" (Ph.D. diss., University of Wisconsin–Madison, 1975), 304. Historians' estimates of the cost of purchasing and installing sound systems vary widely; this is due partly to the fact that prices varied according to the quality of equipment purchased. According to *Overture,* sound systems and installation costs ranged from $6,500 to $25,000. *Overture,* June 1928, 19. Swenson maintains that costs ranged from $16,000 to $25,000. Swenson, "Entrepreneur's Role," 411. As a result of sound films, net profits for producers went up 400 percent between 1927 and 1929. Sklar notes that much of the savings theater owners gained by cutting live talent went to producers, who demanded higher rental rates for sound movies. Sklar, *Movie-Made America,* 152.

31. *Film Daily,* October 28, 1929, 1; Robert Sobel, *RCA* (New York: Stein and Day, 1986), 80. *Film Daily* reported that thirty-five hundred theaters had been wired for sound by 1930. Sobel states that by early 1929 four thousand theaters had been wired or were in the process of conversion. He maintains that by 1929 theaters had spent $37 million on sound equipment. See also Sklar, *Movie-Made America,* 152; Garth Jowett, *Film: The Democratic Art* (Boston: Little, Brown and Company, 1976), 475; and Balio, *American Film Industry,* 113.

32. Exhibitors promised that sound films would evoke "the highest emotional expression." "In the history of motion pictures," one advertisement for *Don Juan* explained, "no production has ever so combined the qualities of entertainment and box office power." See Carl Laemmle, "The Business of Motion Pictures," in Balio, *American Film Industry,* 162–66; *Motion Picture Classics,* August 1929, 6; October 1929, 99; and *Exhibitor's Herald,* February 12, 1927, 14–15.

33. *Film Daily,* October 28, 1929, 1; Sklar, *Movie-Made America,* 152.

34. [Chicago] *Federation News,* September 4, 1926, 7; September 18, 1926, 1–2. The wages of bandleaders in Chicago theaters were often one and a half times higher than the wages of sidemen.

35. Ibid., September 15, 1928, 14.

36. Interview with John TeGroen, former president of Los Angeles Local 47 of the AFM, Los Angeles, March 7, 1987. TeGroen remembered that when AFM president Weber heard new electrical phonographic recordings in "about 1924," he reportedly said, "That's beautiful, but it will never replace live musicians." See also *Overture*, January 1928, 18–19; and Behlmer, "Tumult, Battle, and Blaze," 45.

37. *Overture*, August 1927, 1; June 1928, 19; November 1928, 18.

38. *Official Proceedings*, 1929, 53; 1930, 30.

39. *Overture*, February 1929, 2; *Official Proceedings*, 1929, 53; *Monthly Labor Review* 27 (November 1928): 1029–30.

40. Gomery, "Coming of Sound," 299; *Exhibitor's Herald*, September 15, 1928, 19; October 27, 1928, 26; *Motion Picture News*, September 8, 1928, 785.

41. *Exhibitor's Herald*, October 27, 1928, 26. Estimates of the number of theaters in Chicago in 1928 range from 250 to nearly 400, depending on the source consulted and, apparently, on the definition of the metropolitan area used. *Federation News*, September 6, 1928, 3; *Los Angeles Citizen*, September 7, 1928. See also Cohen, *Making a New Deal*, 130.

42. Gomery, "Coming of Sound," 302–4; *Exhibitor's Herald*, September 15, 1928, 58; *Motion Picture News*, September 15, 1928, 845.

43. *Overture*, June 1929, 7; *Official Proceedings*, 1930, 32.

44. *Overture*, June 1929, 13; Leiter, *Musicians and Petrillo*, 60.

45. *Los Angeles Citizen*, August 3, 1928.

46. *Federation News*, July 14, 1928, 6.

47. *American Federationist*, September 1931, 1070; *Film Daily*, October 24, 1929, 10; *Official Proceedings*, 1931, 36–37; 1930, 32; *Monthly Labor Review* 27 (November 1928): 1029–30; *Federation News*, November 9, 1929. Ironically, a union suffering from a communications revolution utilized mass media to communicate its problem to the public and enlist public help.

48. *Exhibitor's Herald*, September 7, 1929, 29; September 21, 1929, 36; *Film Daily*, November 15, 1929, 7; December 2, 1929, 7; August 28, 1930, 1; September 14, 1930, 1; September 19, 1930, 1.

49. *Film Daily*, September 12, 1930, 1; December 22, 1930, 1; October 3, 1930, 3.

50. Balio, *American Film Industry*, 216–20; Jowett, *Film: The Democratic Art*, 473–75.

51. Balio, *American Film Industry*, 216–20; *Film Daily*, October 3, 1930, 3.

52. Cohen, *Making a New Deal*, 128–30. All of Chicago's fifty largest theaters apparently featured sound movies. See also Sklar, *Movie-Made America*, 162; Balio, *American Film Industry*, 213; and Jowett, *Film: The Democratic Art*, 482.

53. *Overture*, December 1929, 2; *Official Proceedings*, 1930, 28; "Machines Threaten Music," clipping in Bagley Collection; *Monthly Labor Review* 33 (November 1931): 1015–17; *Film Daily*, October 28, 1929, 1; *Monthly Labor Review* 33 (August 1931): 261.

54. *Monthly Labor Review* 33 (August 1931): 261; Leiter, *Musicians and Petrillo*, 60; *Overture*, April 1938, 18. In Los Angeles, a city with a population of 1.5 million and the nation's third largest musician's union, only thirty musicians worked in theaters in 1938.

55. *Federation News*, August 30, 1930, 79. Hayden blasted the use of "robot music" and blamed big business for the musicians' troubles: "We know that the forces behind

the moving picture trust are Bell Telephone, the Westinghouse, the General Electric, the Radio Corporation, the DuPonts, and so forth."

56. *Monthly Labor Review* 33 (November 1931): 1011–16; *Film Daily,* September 11, 1930, 1.

57. *Monthly Labor Review* 33 (November 1931): 1014–16. One former organist had become chief usher and advertising agent for the theater in which he once performed. He made $25 a week, far less than the $62 he had earned, and the loss had forced him to move in with his mother-in-law. One musician who had earned $40 a week was "absolutely without means except for the little income that his wife [brought] in by occasional sewing."

58. *International Musician,* December 1938, 15. The AFM reported that Nicholas Schenck, representing both Metro-Goldwyn-Mayer and 20th Century–Fox, made this comment in 1938 labor negotiations.

59. *Overture,* May 1929, 21; *American Federationist,* October 1928, 1183. Weber said in 1928 that the musician "is not professionally jealous of this machine . . . because he does not concede the possibility that canned music can ever attain artistic quality, but he is fearful that it may work at least a temporary revolution in the musical field."

Chapter Three: Encountering Records and Radio

1. *Billboard,* November 29, 1941, 90–91; Russell Sanjek, *American Popular Music and Its Business: The First Four Hundred Years* (New York: Oxford University Press, 1988), 3:1–15; George N. Gordon, *The Communications Revolution: A History of Mass Media in the United States* (New York: Hastings House, 1977), 165–68.

2. *Billboard,* January 3, 1942, 70–71.

3. Ibid., 71; Sanjek, *American Popular Music,* 2:381. Charles K. Harris's ballad "After the Ball" sold five million copies in the 1890s. Lewis A. Erenberg, *Steppin' Out: New York Nightlife and the Transformation of American Culture* (Chicago: University of Chicago Press, 1981), 72; Craig H. Roell, *The Piano in America, 1890–1940* (Chapel Hill: University of North Carolina Press, 1989), 49–50.

4. Roell, *Piano in America,* 52–59; Neil Leonard, *Jazz and the White Americans* (Chicago: University of Chicago Press, 1962), 97; Arnold Shaw, *The Jazz Age: Popular Music in the 1920's* (New York: Oxford University Press, 1987), 70–84.

5. Leonard, *Jazz and the White Americans,* 96–97.

6. John Harvith and Susan Edwards Harvith, eds., *Edison, Musicians, and the Phonograph: A Century in Retrospect* (New York: Greenwood, 1987), 25–29, 47–50.

7. Ibid., 48–50; Sanjek, *American Popular Music,* 2:391. The total money Caruso earned from Victor was actually much higher. He was apparently paid an additional $10,000 for a second ten sides and then earned royalties over the next several years far exceeding these payments.

8. *Official Proceedings,* 1925–26, 33. In 1925–26 the union's proceedings were biennial. The fidelity of early recordings was so poor that the emphasis was on singers, not instrumentalists. Orchestral recordings did not become technologically practical until 1913. Vern L. Countryman, "The Organized Musicians" (part 2), *University of Chicago Law Review* 16 (Winter 1948): 240–50.

9. *Billboard,* November 29, 1941, 91.

10. Lawrence W. Lichty and Malachi C. Topping, eds., *American Broadcasting* (New York: Hastings House, 1975), 23–25; Erik Barnouw, *A Tower in Babel* (New York: Oxford University Press, 1968), 19–20, 130–31.

11. In 1922, in an effort to protect the signals of ships at sea from powerful new transmitters on land, the department assigned a higher wavelength to land-based broadcasters. Hoover warned radio that excessive use of "mechanical music" might dampen public interest in the new industry. Countryman, "Organized Musicians," 242; *Official Proceedings,* 1955, 218–19. See also Susan J. Douglas, *Inventing American Broadcasting, 1899–1922* (Baltimore: Johns Hopkins University Press, 1987), 315–16; and Barnouw, *Tower in Babel,* 121.

12. *New York Times,* February 3, 1924, 8–9.

13. Memoirs of Dorothy Humphreys, Dorothy Stephens Humphreys Papers, box 1, folder 5, Ohio Historical Society, Columbus.

14. Ibid. WCAH became WBNS. Humphreys performed for the station for twenty-five years.

15. David R. Mackey, "The Development of the National Association of Broadcasters," *Journal of Broadcasting* 1 (Fall 1957): 307–8. Mackey notes that in 1922, sales reports of Victor Records were only 50 percent of what management had expected. Barnouw, *Tower in Babel,* 119; *Variety,* August 18, 1922, 1.

16. *New York Times,* February 3, 1924, 8.

17. Barnouw, *Tower in Babel,* 134–35; *New Majority,* April 5, 1924, 3; *Intermezzo,* April 1924, 2.

18. *Intermezzo,* April 1924, 2.

19. *Overture,* May 1933, 6–9; Philip K. Eberly, *Music in the Air: America's Changing Tastes in Popular Music, 1920–1980* (New York: Hastings House, 1982), 14–23.

20. In 1934 the wages of musicians in radio could range from $30 to $100 a week and more. Musicians in radio worked twelve to forty hours a week. *Official Proceedings,* 1934, 37. Also *Overture,* May 1933, 7.

21. John W. Spaulding, "1928: Radio Becomes a Mass Advertising Medium," in Lichty and Topping, *American Broadcasting,* 225; Edgar E. Willis, *Foundations in Broadcasting: Radio and Television* (New York: Oxford University Press, 1951), 53; Barnouw, *Tower in Babel,* 133.

22. In 1928, broadcasters collected over $10 million from national and local advertisers. Spaulding, "1928," 223–24. Radio transmitters also became far more reliable in terms of sound quality over longer distances. Eberly, *Music in the Air,* 34; *New York Times,* May 13, 1928.

23. *Official Proceedings,* 1934, 33; 1935, 49; Robert Sobel, *RCA* (New York: Stein and Day, 1986), 14. Station WHAS in Louisville, Kentucky, set up remote studios in the local music stores to gain access to musical talent. See also *Federation News,* June 28, 1928, 3.

24. *Federation News,* January 1947, 5. By 1947 this trade journal had switched to monthly publication. Also Willis, *Foundations in Broadcasting,* 38.

25. Abel Green and Joe Laurie, Jr., *Show Biz: From Vaude to Video* (New York: Henry Holt and Company, 1951), 134; Leonard, *Jazz and the White Americans,* 106.

26. *Official Proceedings,* 1925–26, 33–35.

27. J. Fred MacDonald, *Don't Touch That Dial! Radio Programming in American*

Life, 1920–1960 (Chicago: Nelson-Hall, 1979), 25. WEAF later became WRCA and eventually WNBC. WJZ was originally located in Newark, New Jersey, and relocated to New York. Some network affiliates hooked up to both WEAF and WJZ. RCA purchased WEAF from AT&T. Fees for radio programs depended partly on the size of the audiences to whom affiliates broadcast.

28. Michele Hilmes, *Hollywood and Broadcasting: From Radio to Cable* (Urbana: University of Illinois Press, 1990), 51. Willis suggests that the names Red and Blue Network stemmed from the different-colored wires at WEAF and WJZ. Willis, *Foundations in Broadcasting*, 39.

29. Willis, *Foundations in Broadcasting*, 28–30; Sobel, *RCA*, 102–71.

30. Tom Lewis, *Empire of the Air: The Men Who Made Radio* (New York: Harper-Collins, 1991), 182–83; Hilmes, *Hollywood and Broadcasting*, 51; Sobel, *RCA*, 25, 46; Willis, *Foundations in Broadcasting*, 40–41. At age twenty-seven Paley purchased CBS after Columbia Phonograph Record Company formed Columbia Phonograph Broadcasting System. Paley's family, which owned the Congress Cigar Company, had used network advertising with impressive results.

31. *Official Proceedings*, 1938, 122, 155; Sobel, *RCA*, 25, 46. Some of the stations associated with the Mutual Network, such as KHJ in Los Angeles, controlled regional networks of their own. KHJ served as the flagship station for the Don Lee Network. The *Chicago Tribune* and R. H. Macy Company apparently exerted considerable financial control over the Mutual Network. C. Joseph Pusateri, *A History of American Business* (Arlington Heights, Ill.: Harlan Davidson, 1988), 272.

32. *Billboard*, October 18, 1941, 5; January 17, 1942, 6; Sanjek, *American Popular Music*, 3:256. The origins of ABC can be traced to a legal suit Mutual filed against RCA for $10.275 million in damages. The action was taken under the Sherman Antitrust Act. A few regional networks, such as the "Dixie," "Yankee," and "Southwest" chains, connected dozens of other stations. Pusateri, *History of American Business*, 268–70; *Official Proceedings*, 1934, 33.

33. Willis, *Foundations in Broadcasting*, 17–50; Sobel, *RCA*, 102–71.

34. *Variety*, March 11, 1931, 68; Robert Dupuis, *Bunny Berigan: Elusive Legend of Jazz* (Baton Rouge: Louisiana State University Press, 1993), 45–49.

35. *Monthly Labor Review*, August 1931, 3; "What Price Recording?" *Overture*, April 1938, 18; *International Musician*, November 1937, 16. Commercial programs initially relied on staff orchestras, but in an attempt to spread job opportunities, musicians' unions began prohibiting staff musicians from offering services to sponsors of commercial programs. Federal Communications Commission, "Early History of Network Broadcasting (1923–26) and the National Broadcasting Company," in Lichty and Topping, *American Broadcasting*, 171.

36. Shaw, *Jazz Age*, 175.

37. George T. Simon, *The Big Bands*, 4th ed. (New York: Schirmer Books, 1981), 58. The networks occasionally made remotes during early-evening hours as well.

38. As quoted in ibid., xiii, 16–22. Sinatra wrote the foreword to the book. Also "Traveling Bands Working in This Jurisdiction, January 1937," Memo, Papers of Local 47. On musicians' travel experiences, see also David W. Stowe, "Jazz in the West: Cultural Frontier and Region during the Swing Era," *Western Historical Quarterly* 23 (February 1992): 65–66.

39. As quoted in Simon, *Big Bands,* 33–39. On vocalists, see also Eberly, *Music in the Air,* 108–13.

40. As quoted in Stowe, "Jazz in the West," 61–62. See also Nat Shapiro and Nat Henthoff, eds., *Hear Me Talkin' to Ya: The Story of Jazz as Told by the Men Who Made It* (New York: Dover, 1966), 39, 43; and Garvin Bushell, as told to Mark Tucker, *Jazz from the Beginning* (Ann Arbor: University of Michigan Press, 1988), 326–27.

41. *Official Proceedings,* 1932, 232, 250. Philadelphians attempted to levy a $3 tariff on stations for each network program they received. If the station employed a staff orchestra, the tariff would be lower.

42. *Federation News,* July 17, 1926, 3; Countryman, "Organized Musicians," 248–49. Perplexing court rulings about the role of government in broadcasting and harsh criticism from broadcasters opposed to efforts to regulate radio caused confusion at the FRC. As a result, officials requested an opinion concerning federal authority in radio from the attorney general, who stated that Congress had "broken down" the regulatory power of the FCC over radio by not passing the necessary legislation. This statement apparently prompted the FRC to abandon policies established by Hoover.

43. *Billboard,* January 3, 1942, 67, 73; Leonard, *Jazz and the White Americans,* 91.

44. *Billboard,* November 29, 1941, 91; Eberly, *Music in the Air,* 86. Albums were also introduced in 1935. Jukebox operators paid only 21 cents for Decca records. See also James Lincoln Collier, *Benny Goodman and the Swing Era* (New York: Oxford University Press, 1989), 188–94.

45. *Billboard,* October 4, 1941, 65; *Overture,* August 1942, 22.

46. *Official Proceedings,* 1932, 250; Eberly, *Music in the Air,* 76–79, 178–94. The first commercial use of transcriptions by radio was apparently in 1929, when WOR demonstrated their capacity for high-quality reproduction of sound. Transcriptions spun at a slow thirty-three and one-half revolutions per minute.

47. In 1938, 46 percent of RCA-Thesaurus clients were not affiliated with any network. *Broadcasting,* April 15, 1939, 64. Also FCC, "Early History of Network Broadcasting," 171; Countryman, "Organized Musicians," 251; and Eberly, *Music in the Air,* 134.

48. *Official Proceedings,* 1934, 34; *Intermezzo,* August 1939, 3; Countryman, "Organized Musicians," 250. Neither bandleaders nor sidemen received royalties on the sales of transcriptions, since they signed "work for hire" contracts that prohibited future royalty claims. Sanjek, *American Popular Music,* 3:130.

49. Joseph G. Rayback, *A History of American Labor* (New York: Macmillan, 1966), 213–15; Ronald L. Filippelli, *Labor in the USA: A History* (New York: Random House, 1984), 152–53.

50. Barnouw, *Tower in Babel,* 121–22; Mackey, "Development of NAB," 310–17; *Variety,* May 24, 1923, 40. The ability of NAB to invade labor's domain was made clear in 1940, when the association organized Broadcast Music Incorporated (BMI), a competitor to ASCAP made up of songwriters who agreed to abandon royalty demands in return for the chance to have their music on radio. With the formation of BMI, broadcasters had two sources from which they pulled musical selections. See also John Kenneth Galbraith, *American Capitalism: The Concept of Countervailing Power* (Boston: Houghton Mifflin, 1956), 113.

51. Mackey, "Development of NAB," 305–6. Mackey reports that in 1956 the NAB

maintained a staff of seventy-seven in its four-story building in Washington, sent out over half a million pieces of mail, and had over twenty-five hundred broadcasters and associates attend its annual convention. See also Barnouw, *Tower in Babel*, 121–22. David Noble, *America by Design: Science, Technology, and the Rise of Corporate Capitalism* (New York: Oxford University Press, 1977), explains more thoroughly how industry defines business development as progress.

52. As quoted in Sally Bedell Smith, *In All His Glory: The Life of William S. Paley, the Legendary Tycoon, and His Brilliant Circle* (New York: Simon and Schuster, 1990), 41. Paley's father had moved from Chicago largely because of the rising power of the local cigarmakers' union. During the 1919 strike young Paley ignored requests of the local Cigar Manufacturers' Association as well as the cigarmakers' union. See also Patricia A. Cooper, *Once a Cigar Maker: Men, Women, and Work Culture in American Cigar Factories, 1900–1919* (Urbana: University of Illinois Press, 1987), 301.

53. *Official Proceedings*, 1933, 252.

54. Ibid., 1930, 193; Robert D. Leiter, *The Musicians and Petrillo* (New York: Bookman Associates, 1953), 60; *Official Proceedings*, 1933, 193.

55. *Official Proceedings*, 1936, 205; 1934, 213.

56. The proposed tariff was $5. If the station receiving a chain program employed a staff orchestra, the tariff would be lowered to $3. Ibid., 1932, 251; 1935, 45–46.

57. Anders S. Lunde, "The American Federation of Musicians and the Recording Ban," *Public Opinion Quarterly* 6 (Spring 1948): 47.

58. *Official Proceedings*, 1928, 7–16; 1929, 53; 1934, 7–15, 249; 1935, 45–47. To locals that proposed that transcriptions be denied stations on the federation's "unfair list," president Weber explained that the opportunity for enforcing such a rule would be "nil." To locals that wanted to prevent networks from sending chain programs to stations not employing a staff orchestra, Weber maintained that such action would also be unenforceable and would involve the union in a dangerous altercation with the FCC.

59. "What Price Recording?" 18.

60. The jobs of nearly one thousand musicians working for the three major networks had become an important asset to the financially troubled union. *Official Proceedings*, 1935, 46; 1937, 92.

61. Joseph N. Weber to All Local Unions, November 6, 1933, Papers of Local 47.

62. Postal telegraph from Weber to F. D. Pendleton, president of AFM Local 47 of Los Angeles, December 23, 1933, and Weber to "all locals of the AFM," November 6 and 18, 1933, Papers of Local 47. Weber hoped to create part-time employment totaling 25 percent of the hours worked by regularly employed instrumentalists. Also *Intermezzo*, June 1931, 1; *Official Proceedings*, 1934, 31; C. Ellsworth Wylie, general manager of KHJ, to Weber, January 12, 1934, Papers of Local 47.

63. Gerald King, manager, KFWB, to Frank Pendleton, president of AFM Local 47 of Los Angeles, Papers of Local 47. Clearly, relations between the AFM and station KFWB were not amicable. The manager of KFWB wrote, "Mr. Weber states he has no objection to our signing a contract with a musician if we will agree that his organization can make any changes in the contract which he may desire at any time hereafter." The AFM, the manager believed, is "willing to trade a horseshoe for a horse any time." See also *Official Proceedings*, 1934, 80.

64. *Official Proceedings*, 1937, 57. For a legalistic account of the formation of NAPA, see Countryman, "Organized Musicians," 253–63. Also Leiter, *Musicians and Petrillo,* 67.

65. *Official Proceedings*, 1937, 57.

66. *International Musician*, January 1938, 1, 4; Countryman, "Organized Musicians," 255–56.

67. Countryman, "Organized Musicians," 256.

Chapter Four: Playing in Hollywood between the Wars

1. Local 47 president Frank D. Pendleton to Joseph N. Weber, November 27, 1933, Papers of Local 47.

2. *Overture*, July 1938, 2; September 1938, 2; May 1933, 6–9. Filmmaking was the leading industry in Los Angeles in the 1920s. In 1939 it employed thirty thousand to forty thousand workers and spent an estimated $190 million on the manufacture of motion pictures. The city's radio orchestras increased in size and number as the emergence of "sound stars" encouraged the networks to broadcast more shows from Los Angeles. In 1938 six Hollywood radio stations paid over $5 million to six hundred film stars. Carey McWilliams, *Southern California: An Island on the Land* (Santa Barbara: Peregrine Smith, 1973), 339–40.

3. Further solidifying the AFM's grip on musical services was the fact that employers of musicians, in constant search of larger audiences, competed for the best bands and orchestras, not simply the least expensive. *Constitution and By-laws of the MMPA of Los Angeles*, 1937, in Bagley Collection. Also letters from union officials to local musical groups, 1934–41, Papers of Local 47.

4. Everett Lee Refior, "The American Federation of Musicians: Organization, Policies, and Practices" (master's thesis, University of Chicago, 1955), 172; David Bordwell, Janet Staiger, and Kristin Thompson, *The Classical Hollywood Cinema: Film Style and Mode of Production to 1960* (New York: Columbia University Press, 1985), 312. The joint committee consisted of five producer representatives and five union representatives. Also J. W. Buzzell to Local 47 board of directors, August 11, 1936, Papers of Local 47. Three strong associations that often opposed the interests of labor, especially in the manufacturing sector in Los Angeles, were the Merchants and Manufacturers Association, the Chamber of Commerce, and the Downtown Retail Merchants' Association.

5. C. L. Bagley, "Seventy-five Years of Growth," in special edition of *Overture*, 1925, 39–40; Louis B. Perry and Richard S. Perry, *A History of the Los Angeles Labor Movement* (Berkeley: University of California Press, 1963), 489. Also Robert M. Fogelson, *The Fragmented Metropolis: Los Angeles, 1850–1930* (Berkeley: University of California Press, 1967), 78; Leo C. Rosten, *Hollywood: The Movie Colony, the Movie Makers* (New York: Harcourt, Brace and Company, 1941), 371; and Arthur C. Verge, *Paradise Transformed: Los Angeles during the Second World War* (Dubuque, Iowa: Kendall/Hunt, 1993), 1–7.

6. *Official Proceedings*, 1919, 8; 1920, 8; 1930, 8; 1940, 8. See also Perry and Perry, *History of Los Angeles Labor*, 489; Robert D. Leiter, *The Musicians and Petrillo* (New York: Bookman Associates, 1953), 69, 80–83, 95; *Overture*, April 1928, 2; August 1950,

34; and Murray Ross, *Stars and Strikes: The Unionization of Hollywood* (New York: Columbia University Press, 1941). Trends in the membership of other locals of the AFM underscored the fact that musicians were looking to Los Angeles as a place to further their careers. In 1929 Philadelphia had a population 50 percent larger than Los Angeles but nearly eight hundred fewer union musicians. Louisville, Kentucky, was one-quarter the size of Los Angeles, but its musicians' union had only one-tenth as many members as Local 47. In the 1940s, when membership in the Los Angeles union mushroomed to over thirteen thousand, the federation's fourth largest branch, in Detroit, had fewer than five thousand members. By that time Local 47 had surpassed Chicago as the second largest affiliate of the AFM.

7. *Official Proceedings*, 1929, 18; interview with John TeGroen, Los Angeles, October 1, 1988.

8. TeGroen interview, October 1, 1988; interview with Phil Fischer, radio representative of Local 47, Los Angeles, October 1, 1988.

9. Contractors were in constant competition with one another for the services of "name" musicians. Interviews with Eudice Shapiro, Los Angeles, January 14, 15, 1989. In addition to the major studios, a dozen or more independent companies used live, though smaller, orchestras. *Overture*, April 1938, 18–19; interview with Phil Fischer, Los Angeles, April 3, 1987.

10. Robert R. Faulkner, *Hollywood Studio Musicians: Their Work and Careers in the Recording Industry* (Chicago: Aldine and Atherton, 1971), 144–47; interview with Al Hendrickson, Los Angeles, June 3, 1994.

11. *Overture*, January 1942, 21; interview with Phil Fischer, Los Angeles, January 19, 1989. Union rules required individuals in leadership positions to receive at least "double scale." The length of the workday of film musicians varied according to rehearsal schedules and production problems, but union rules demanded extra pay for longer hours. Smith got his break when a contractor, acting on the advice of another musician, requested his services. Interviews with Art Smith, Los Angeles, September 25, November 21, 1988.

12. Shapiro interviews, January 14, 15, 1989. Concertmasters were common in symphony orchestras as well as in motion-picture orchestras. Concertmasters not only performed in film orchestras but helped string players orchestrate common fingerings for special musical passages and checked the bowing and tuning of instruments.

13. Jas Obrecht, "Pro's Reply: Al Hendrickson—The Way We Were," *Guitar Player*, March 1988, 18; interview with Al Hendrickson, Los Angeles, January 21, 1989. Other instrumentalists have echoed Hendrickson's comments regarding the demands on sight-reading skills. Woodwind specialist Art Smith explained that even the most complex music was not handed out in advance: "They assumed you could play it. At times the pages looked like wall-to-wall notes." Smith interview, September 25, 1988. Violinist Louis Kaufman recalled, "You never knew what they would throw at you." John Harvith and Susan Edwards Harvith, eds., *Edison, Musicians, and the Phonograph: A Century in Retrospect* (New York: Greenwood, 1987), 124.

14. Faulkner, *Hollywood Studio Musicians*, 108, 97–106.

15. Harvith and Harvith, *Edison*, 108–30. Kaufman moved to Los Angeles from Portland, Oregon, and performed solos in perhaps five hundred films during his twenty-five-year career in the studios. Kaufman noted that film musicians worked

much earlier hours than instrumentalists in traditional areas of employment: "Usually we would start at nine or ten o'clock in the morning, have a break for lunch, and work during the day until five, six, or seven." Since instrumental solos were generally recorded after the full orchestra was dismissed, Kaufman occasionally worked late. When Local 47 insisted on higher wages for work done after midnight "things began to be more regularized. From then on," he said, "the sessions ended promptly at midnight, if we were still working at that hour." He said that the work of soloists "was a great strain and not everybody was able to take it."

16. Charles Hofmann, *Sounds for the Silents* (New York: DBS, 1970), 38–41. Actress Lillian Gish agreed that D. W. Griffith did not use sideline musicians. "Griffith would never allow any music on the set. So many of us loved music," Gish explained, "if it was played as we tried to work we would stop to listen to it and never get the film finished."

17. Quotation from *Overture,* March 1948, 15. Local 47 established detailed guidelines for the activities of sideliners. Rules even took into account changing weather conditions and costumes worn. Also interview with John TeGroen, Los Angeles, December 29, 1987.

18. Ernest Gold, "My First Movie Score," *Film Music Notes* 5 (November 1945): 7.

19. Lawrence Morton, "Film Music Profile: Adolph Deutsch," *Film Music Notes* 9 (November–December 1949): 4; Morton, "Film Music Profile: André Previn," *Film Music Notes* 10 (January–February 1951): 4. Film composer Herbert Stothart explained that his job was to use the orchestra so that the audience would not become "music conscious at the expense of the dialogue or drama." Herbert Stothart, "The Orchestra—Hollywood's Most Versatile Actor," *Film Music Notes* 3 (February 1944): 14. Also Harvith and Harvith, *Edison,* 121–25.

20. An eighty-piece orchestra recorded the music for *King Kong.* The music, for which Steiner won an Oscar, increased the film's budget by $50,000. William LeBaron was the RKO production chief who hired Steiner, but Merian C. Cooper produced *King Kong.* See Orville Goldner and George E. Turner, *The Making of King Kong: The Story behind a Film Classic* (New York: A. S. Barnes and Company, 1975), 190–91; Christopher Palmer, *The Composer in Hollywood* (London: Marion Boyars, 1990), 27–29; and *King Kong* (1933), Wong Audio-Visual Center, University of Hawaii at Manoa, Honolulu.

21. *King of Jazz* (1930), Wong Audio-Visual Center, University of Hawaii at Manoa, Honolulu. Carl Laemmle, Jr., produced this movie, with orchestration by Ferde Grofe and Milton Ager. Featured songs included "It Happened in Monterey," "Rhapsody in Blue," "Happy Feet," "Ragamuffin Romeo," "A Bench in the Park," "Melting Pot," and "The Song of the Dawn."

22. Fischer interview, April 3, 1987.

23. In an effort to increase the number of radio jobs, Local 47 prohibited musicians from working in both sustaining and commercial orchestras. Most locals did not make such sharp distinctions. Ibid.; Philip K. Eberly, *Music in the Air: America's Changing Tastes in Popular Music, 1920–1980* (New York: Hastings House, 1982), 23; interview with Lenny Atkins, Los Angeles, September 7, 1988. Throughout the 1930s both large and small stations employed sustaining bands. They also relied on the services of part-time musicians when orchestras were expanded for special programs.

24. Interview with Lenny Atkins, Los Angeles, November 18, 1988; Smith interviews, September 25, November 21, 1988.

25. *Overture*, May 1933, 6–9; January 1943, 9; Fischer interview, April 3, 1987; interview with Henry Gruen, Los Angeles, November 14, 1988; Atkins interview, November 14, 1988. Until 1940, when the AFM gave locals greater control over radio because of the threat of antitrust action against the union, the size of network orchestras and the annual wages of their members were largely determined in labor negotiations between broadcasters and the AFM's executive board. *Broadcasting*, March 1, 1939, 64. The minimum-size-orchestra policy was not followed in the motion-picture industry until 1944, when film companies agreed to thirty-six- to fifty-piece orchestras hired on an annual basis. Leiter, *Musicians and Petrillo*, 82–83. In 1945, network and local radio stations paid approximately $12 million in wages to twenty-nine hundred staff musicians. *Official Proceedings*, 1946–47, 61.

26. With a smooth southern accent and a talent for comedy, the popular bandleader made his musical quiz show, *The Kay Kyser Kollege of Musical Knowledge*, one of radio's highest-rated programs. *Overture*, April 1943, 3. In 1945, musicians working on commercial programs in Los Angeles earned approximately $3 million in total wages. *Official Proceedings*, 1946–47, 62.

27. Gruen interview, November 14, 1988; Fischer interview, April 3, 1987. Unlike many musicians, Gruen never auditioned for the job. His personal connections as well as his reputation as a skilled and dependable horn player had secured studio employment. "I got it," he said, "because I had friends in the band."

28. *Overture*, May 1933, 6–9; January 1943, 9.

29. Interviews with Jack Bunch, ABC bandleader in 1942, and Roc Hilman, radio bandleader at station KHJ, Los Angeles, November 15, 1988.

30. Edgar E. Willis, *Foundations in Broadcasting: Radio and Television* (New York: Oxford University Press, 1951), 290–92.

31. Ibid., 156–57.

32. Smith interviews, September 25, November 21, 1988. Also Susan Smulyan, *Selling Radio: The Commercialization of American Broadcasting, 1920–1934* (Washington, D.C.: Smithsonian Institution Press, 1994), 7; and Steven Loza, *Barrio Rhythm: Mexican American Music in Los Angeles* (Urbana: University of Illinois Press, 1993), 33. Los Madrugadores means "the early risers."

33. *Overture*, 1956, 6; *Official Proceedings*, 1932, 16; interview with John TeGroen, Los Angeles, February 3, 1989. On discrimination and segregation in Los Angeles during the 1930s and 1940s, see Verge, *Paradise Transformed*, 39–60.

34. At AFM conventions, representatives of all-black locals often stayed at segregated hotels and faced various forms of discrimination. See Donald Spivey, *Union and the Black Musician: The Narrative of William Everett Samuels and Chicago Local 208* (New York: University Press of America, 1984), 56–57.

35. Loza, *Barrio Rhythm*, 21, 33. In the 1940s Los Angeles became a center for small independent companies specializing in blues recordings. Among those companies were Aladdin Records, Exclusive/Excelsior, Black and White Records, and 4 Star Records. When the companies were founded is unclear. See William Barlow, *Looking up at Down: The Emergence of Blues Culture* (Philadelphia: Temple University Press, 1989), 33–34.

36. Across the nation, local unions vehemently opposed the use of transcriptions, since these long-playing records could be used as substitutes for live radio bands. Union opposition to transcriptions offers another example of how labor, through resistance to technological change, could shape industrial practice. Federation policies limited the use of transcriptions and encouraged broadcasters to use live orchestras. *Overture,* September 1938, 2. A few musicians worked steadily for record manufacturers and earned wages comparable to those of instrumentalists in radio and motion pictures. Hendrickson interview, January 21, 1989.

37. The influx of accomplished musicians seeking studio jobs made club and hotel work increasingly competitive, especially in the best places. Smith interviews, September 25, November 21, 1988.

38. *Official Proceedings,* 1934, 33; *Overture,* May 1933, 6–9; Eberly, *Music in the Air,* 69; Fischer interview, April 3, 1987; Gruen interview, November 14, 1988.

39. Smith interviews, September 25, November 21, 1988; TeGroen interview, February 3, 1988; Eberly, *Music in the Air,* 96. The Culver City Cotton Club was an imitation of the more famous Harlem nightclub of the same name.

40. Neil Leonard, *Jazz: Myth and Religion* (New York: Oxford University Press, 1987), 71–73; interview with Bob Fleming, Honolulu, August 16, 1993.

41. Al Hendrickson to author, June 8, 1994.

42. Michele Hilmes, *Hollywood and Broadcasting: From Radio to Cable* (Urbana: University of Illinois Press, 1990), 62–63.

43. Proposed Yearly Quota System, March 1, 1955, Bulletin, Papers of Local 47. This eleven-page bulletin gives a brief history of the quota laws, followed by ten pages detailing the various quota regulations.

44. Cliff Webster to Joe Weber, June 25, 1936, Papers of Local 47. Beginning in 1934 the AFM required traveling bands to charge employers 10 percent above union price lists for performances, though the Ten Percent Law did not stop the influx of outside bands into Los Angeles. This law, which superseded the Thirty Percent Law, applied to union engagements of a week or longer. Joseph N. Weber to Local Unions, February 26, 1935, Papers of Local 47.

45. Webster to Weber, June 25, 1936, Papers of Local 47.

46. Joe Weber to F. D. Pendleton, July 7, 1936, Radiogram, Papers of Local 47; Harry E. Brenton, secretary treasurer of AFM, to Local Unions, August 1, 1936, Papers of Local 47.

47. *Overture,* January 1942, 21; TeGroen interview, February 3, 1988; interview with Phil Fischer, Los Angeles, February 3, 1988; interview with John TeGroen, Los Angeles, January 21, 1989.

Chapter Five: Rising Militancy

1. *Official Proceedings,* 1938, 104; *Variety,* August 4, 1937, 30; August 18, 1937, 43.

2. *Billboard,* January 3, 1942, 70–72; *Overture,* August 1942, 22; *Variety,* December 6, 1939, 39; *Official Proceedings,* 1938, 168; 1942, 279. Jukeboxes cost $175 and up; operators typically collected about $8 a week per machine, of which the proprietor received about 20 percent. Operators also paid license fees of $10 a month to the American So-

ciety of Composers, Authors, and Publishers (ASCAP), and usually a small fee of about $1 to local government. Musical Instrumental Distributing Company was a leading distributor of jukeboxes.

3. *Official Proceedings*, 1942, 278–79.

4. *Variety*, November 22, 1939, 39; *Official Proceedings*, 1940, 98; 1938, 95.

5. *Official Proceedings*, 1928, 1–16; 1936, 1–15.

6. *International Musician*, September 1937, 15–19.

7. J. F. McMahon to William Green, July 17, 1937, American Federation of Labor Records, The Samuel Gompers Era, microfilm edition, reel 40, frame 1500, Special Collections Library, University of California, Los Angeles.

8. *Variety*, July 21, 1937, 35.

9. Ibid.; August 18, 1937, 43; *Official Proceedings*, 1938, 167–70; *International Musician*, April 1938, 11.

10. *Variety*, July 28, 1937, 38; *International Musician*, August 1937, 1, 3.

11. Howell John Harris, *The Right to Manage: Industrial Relations Policies of American Business in the 1940s* (Madison: University of Wisconsin Press, 1982), 11, 193.

12. *Official Proceedings*, 1938, 93.

13. Ibid., 94–99; *International Musician*, August 1937, 1, 3. Less controversial propositions would have prohibited manufacturers from recording music without the knowledge of performers, or making new recordings from old ones. One proposal similar to the first suggested that AFM members not offer services at any place that used records unless the place also employed musicians. Another proposal suggested that recordings be given a registered number and those numbers be placed on file with the AFM; the union would then grant permission to use registered recordings under specific conditions. Still another promised to assign a registered number to recordings and then require persons using recordings to obtain permission from the AFM.

14. *Official Proceedings*, 1938, 98–103.

15. Ibid., 120; *Variety*, August 4, 1937, 30.

16. *Official Proceedings*, 1938, 122, 128–30.

17. *Variety*, August 4, 1937, 30; *Official Proceedings*, 1938, 116, 141; *International Musician*, April 1938, 15. Weber had resisted efforts to pinpoint the number of jobs the union hoped to gain, saying only that "a few hundred would be unsatisfactory." At one point he told the network, "I know that you cannot put every unemployed musician to work, but you can help in that direction." Membership of the three-man committee of the executive board rotated regularly.

18. *Official Proceedings*, 1938, 130.

19. *Variety*, August 18, 1937, 43; *International Musician*, November 1937, 1.

20. *Official Proceedings*, 1938, 150–51. William S. Hedges was chairman of IRNA, and seven other station owners served as the organization's principal representatives. Russell Sanjek, *American Popular Music and Its Business: The First Four Hundred Years* (New York: Oxford University Press, 1988), 3:170.

21. *Variety*, September 15, 1937, 35; March 16, 1938, 39; *Official Proceedings*, 1938, 151–59; *International Musician*, November 1937, 1; April 1938, 1. The agreement did not cover the twenty-two smaller stations owned by the major networks; these stations apparently reached individual agreements with the AFM. *Variety* reported that southern stations were most critical of the AFM and were especially opposed to revealing, even

to IRNA, any facts concerning their profits or employment levels. *Variety,* October 6, 1937, 30.

22. *Official Proceedings,* 1938, 158–59. The lowest amount allocated to a single station was $560. Only sixteen stations had annual incomes of less than $15,000 to $20,000.

23. *Variety,* October 6, 1937, 30. The agreement raised NBC's expenditures on staff employment to more than $1 million, and that of CBS to $550,000.

24. *International Musician,* April 1938, 11; October 1938, 3; *Official Proceedings,* 1938, 107, 126, 151, 165, 170–74; *Variety,* March 30, 1938, 31; April 27, 1938, 35. About 40 of the nation's 386 nonaffiliated stations earned 50 percent of this group's income. The union agreed to subtract $15,000 to $20,000 from the annual incomes of many nonaffiliated stations before determining their need for instrumentalists.

25. *Official Proceedings,* 1939, 97; *International Musician,* April 1938, 11; August 1938, 3. Recorders insisted that contract language made them liable for damages and prosecution for secondary boycotts. In regard to dubbing, the union wanted to make sure that musicians who had originally recorded music agreed to having their recorded music altered.

26. *Intermezzo,* September 1938, 1; *Official Proceedings,* 1938, 184. Wage scales in electrical transcription companies were somewhat different. For each fifteen-minute program recorded, musicians received $18. For each thirty-minute program they received $24.

27. *International Musician,* August 1937, 15. The rise of the CIO sparked a new AFM organizing campaign in 1936 and 1937. In 1938 Chicago Local 10 took in announcers and sound effects and production workers at Columbia Broadcasting in response to CIO activity. *Intermezzo,* August 1938.

28. *International Musician,* December 1938, 14. See also Leo C. Rosten, *Hollywood: The Movie Colony, the Movie Makers* (New York: Harcourt, Brace and Company, 1941), 377. Studios apparently owned more than 50 percent of the stock in these fifteen hundred theaters. They owned 25 percent or more of the stock in a total of about twenty-three hundred theaters.

29. *International Musician,* December 1939, 14–15.

30. Ibid.

31. Ellis W. Hawley, *The New Deal and the Problem of Monopoly* (Princeton: Princeton University Press, 1966), 411–55. Under Arnold's direction the number of employees in the Antitrust Division nearly doubled. Arnold organized new subdivisions to receive complaints and carry out investigations. Before teaching at Yale, Arnold had had a wide range of legal and political experience. He had practiced law in Chicago, served in the Wyoming state legislature, been dean of the law school at West Virginia University, and acted as counselor in several New Deal agencies. In 1943 Roosevelt appointed him to the U.S. Court of Appeals for the District of Columbia. See also Michael Conant, "The Paramount Case and Its Legal Background" (1961), in *The Movies in Our Midst: Documents in the Cultural History of Film in America,* ed. Gerald Mast (Chicago: University of Chicago Press, 1982), 594–99; and Ernest Borneman, "United States versus Hollywood: The Case Study of an Antitrust Suit," in *The American Film Industry,* ed. Tino Balio (Madison: University of Wisconsin Press, 1976), 332–45.

32. *International Musician,* December 1938, 15–16; *Official Proceedings,* 1938–39, 78.

33. *Official Proceedings,* 1938–39, 79–81.

34. Ibid.

35. Ibid.

36. Rosten, *Hollywood,* 377; Douglas Gomery, *The Hollywood Studio System* (New York: St. Martin's, 1986), 34, 58, 125, 102, 148, 162, 175; Balio, *American Film Industry,* 332.

37. *Variety,* October 26, 1938, 30.

38. Ibid., October 13, 1937, 28. WBEN vice president A. H. Kirchofer and station director E. H. Twamley wrote the letter. The NAB, they stated, "has failed to meet every crisis that has confronted the industry." They called for reorganization "to save the industry from the dangers that threaten." One force behind the reorganization of the NAB was concern among the networks over the rise of IRNA, an association the networks believed might undermine their position. The networks did not want IRNA to become a permanent dues-paying organization, a development that would reduce their influence.

39. *Broadcasting,* December 1, 1939, 14; November 15, 1939, 20; March 1, 1939, 20.

40. Ibid., November 15, 1939, 20; January 15, 1940, 19. The 1937 National Plan of Settlement was set to expire on January 17, 1940.

41. The following reports from *Broadcasting* give some idea of how fast the profits of the radio and recording industries were rising. The magazine reported that 1935 had been a "record year, . . . showing a 20 percent [increase] over the preceding year," and that 1936 would be better still, with advertising sales hitting $100 million for the first time. From 1941 to 1942 time sales rose by another 14.2 percent. The net profits of RCA, which manufactured radios and records, rose 13 percent from 1940 to 1941. The exact figure for households with radios was 82.8 percent. See *Broadcasting,* January 1, 1936, 10–11; January 27, 1941, 1; February 2, 1942, 1; March 3, 1942, 16; August 3, 1942, 1.

42. Ibid., January 15, 1940, 19.

43. Ibid., December 1, 1939, 1, 14, 76. The head of the Complaint Section of the FCC told Rosenbaum that the Justice Department would act against the union.

44. Ibid., February 1, 1940, 17; *International Musician,* March 1940, 22. Labor contracts between locals and network affiliates covered wages, hours, and conditions of employment, but not the size of staff orchestras, which had been set by the National Plan.

45. Jack Gillette, delegate from Local 47 in Los Angeles, then nominated Petrillo, and Harry Brenton, the AFM's international treasurer, seconded the nomination. Both men praised Petrillo for the "vision" he had shown as head of AFM Local 10. *Intermezzo,* July 1940, 1. See also *Intermezzo,* March 1937, 1–2; April 1937, 1; June 1937, 1; *International Musician,* January 1937, 36; and Anders S. Lunde, "The American Federation of Musicians and the Recording Ban," *Public Opinion Quarterly* 6 (Spring 1948): 47. To maintain employment in theaters, the union insisted that sound-picture recordings made in Chicago (of which there were few) could not be used in theaters charging over 25 cents admission, and even those theaters had to employ at least an organist or pianist. One of Local 10's most controversial requirements was that broadcasters employ a "standby" orchestra of live musicians whenever they played records. The union insisted that standby bands be equal in number to the musicians who per-

formed on any records used. Essentially the standby rule meant that groups of musicians would be paid with no expectation of their rendering services. Since 1911, when the AFM adopted the rule, various locals had demanded that employers hire standby orchestras when employers used nonunion musicians or traveling bands on a temporary basis. Designed to protect and maximize employment, the standby technique resembled tactics practiced by railroad employees, construction workers, and other organized laborers. Employers derisively called such practices "featherbedding." The union had also demanded that Chicago recordings not be used within the territory of other locals without their permission.

46. *Federation News,* September 15, 1928, 5.

47. Interview with Petrillo's son, Lee Petrillo, Chicago, June 26, 1991. Also John Scott Kubach, "Unemployment and the American Federation of Musicians: A Case Study of the Economic Ramifications of Technological Innovations and Concomitant Governmental Policies Relative to the Instrumental Employment Opportunities of the Organized Professional Musicians" (master's thesis, Ohio University, 1957), 70–71. Also Robert D. Leiter, *The Musicians and Petrillo* (New York: Bookman Associates, 1953), 45–47. Petrillo was born on March 16, 1892.

48. *Intermezzo,* July 1932, 1; *Broadcasting,* June 15, 1940, 16.

49. Interviews with Henry Gruen, Los Angeles, October 8, 15, 1993. Gruen received no extra fees when stations made remote broadcasts.

50. *International Musician,* October 1946, 6–7.

51. *Official Proceedings,* 1941, 37–38; *International Musician,* June 1941, 1.

52. *Official Proceedings,* 1941, 38–40.

53. *Interlude* (trade journal of Local 9 of the American Federation of Musicians, Boston), November 1940, 7; *Official Proceedings,* 1941, 38–40; *International Musician,* June 1941, 1, 21.

54. *Official Proceedings,* 1941, 38–40. In some cases band members lost about 15 percent additional wages when the union pulled remotes.

55. *Billboard,* November 29, 1941, 91; November 27, 1941, 145; January 3, 1942, 70–72; *Overture,* August 1942, 22. Roughly 20 percent of all jukeboxes were in three states: Texas, California, and New York. Jukebox interests expressed many of their concerns through the National Automatic Merchandising Association, whose members engaged in the manufacture, placing, and servicing of candy, cigarette, gum, soft drink, nut, and weighing machines.

56. *Billboard,* November 8, 1941, 6; *Official Proceedings,* 1942, 276–78; Carole Elizabeth Scott, "The History of Radio and Television Broadcasting" (unpublished article, West Georgia College, 1990), 23; Sanjek, *American Popular Music,* 3:173.

57. *Official Proceedings,* 1942, 277–78; *International Musician,* June 1937, 3; July 1940, 11; August 1940, 4; September 1940, 19; Sanjek, *American Popular Music,* 3:173.

58. *Official Proceedings,* 1942, 279; *Billboard,* November 8, 1941, 66.

59. *Official Proceedings,* 1942, 326–27. The label fund was also to pay the operator's bookkeeping expenses and the cost of the labels.

60. Ibid., 1944, 56; *International Musician,* February 1943, 1. The new pay raise at the Ringling Brothers' Circus gave the twenty-six white musicians $54 a week. Black musicians got $30.50. Two weeks after the strike began, Ringling agreed to rehire the white musicians at the requested rate, but Petrillo refused the terms: "I told them that

all the men came out when the strike was called and that no settlement could be made unless all the men went back to work."

61. *Billboard,* October 18, 1941, 9; October 25, 1941, 5.

62. Ibid., October 25, 1941, 5; *Official Proceedings,* 1942, 48–50.

63. Robert Dupuis, *Bunny Berigan: Elusive Legend of Jazz* (Baton Rouge: Louisiana State University Press, 1993), 224.

64. *Official Proceedings,* 1942, 27, 54–56. "Exceptions will be made," Petrillo said, "for home consumption, for the armed forces of the United States, and at the request of the President." His promise to allow recordings for home consumption was vague and did not materialize. Petrillo eventually reneged on his promise to change his policy at the request of the president, for by the time Roosevelt appealed to the union to end the ban, the AFM was on the verge of wresting concessions from the networks. Decca Records had already signed a new agreement, and Petrillo insisted that he could not end the ban until all recorders had agreed to the same terms.

65. *International Musician,* July 1942, 1. Petrillo wrote to the recording and transcription companies, "Your license from the American Federation of Musicians for the employment of its members in the making of musical recordings will expire on July 31, 1942, and will not be renewed. From and after August 1, 1942, the members of the American Federation of Musicians will not play or contract for recordings, transcriptions or any other form of mechanical reproductions of music." Petrillo sent a similar letter to all locals of the AFM.

66. *Overture,* August 1942, 22. The Antitrust Division, still under the determined leadership of Arnold, had now filed suit against some 180 business firms and labor unions, a figure that represented almost half of all action taken under the Sherman Act.

67. Ibid.

Chapter Six: Recording Ban

1. Russell Sanjek and David Sanjek, *American Popular Music Business in the Twentieth Century* (New York: Oxford University Press, 1991), 63–65.

2. Interview with Bill Hitchcock, Honolulu, August 13, 1993; interview with Will Brady, Honolulu, August 19, 1993.

3. *Broadcasting,* June 15, 1942, 11–12; Hitchcock interview, August 13, 1993. See also Paul "Fletch" Lindemeyer and Raymond D. Hair, jacket cover of compact disc *Kay Kyser and His Orchestra: The Songs of World War II* (Vintage Records, 1993). The most popular songs in 1943, according to the radio program *Hit Parade,* were "People Will Say We're in Love," "You'll Never Know," and "Paper Doll." Philip K. Eberly, *Music in the Air: America's Changing Tastes in Popular Music, 1920–1980* (New York: Hastings House, 1982), 128–30, 384; Russell Sanjek, *American Popular Music and Its Business: The First Four Hundred Years* (New York: Oxford University Press, 1988), 3:218. Record companies stockpiled both the small records designed for public consumption and long-playing transcriptions used for broadcasting.

4. *Broadcasting,* July 20, 1942, 9. Perhaps the most influential voices in the NAB were those of representatives from RCA, parent company of both NBC and Victor

Records. But owners of CBS, Mutual, and network-affiliated stations also swayed NAB policies. Gleason L. Archer, *Big Business and Radio* (New York: American Historical Company, 1939), 30–32; C. Joseph Pusateri, *A History of American Business* (Arlington Heights, Ill.: Harlan Davidson, 1988), 269–72. RCA's gross income in 1939 was $110 million. Robert D. Leiter, *The Musicians and Petrillo* (New York: Bookman Associates, 1953), 137.

5. *Broadcasting,* July 20, 1942, 7, 54; July 27, 1942, 8; August 17, 1942, 15; August 31, 1942, 10. A poem in the *New York Herald-Tribune* asked Petrillo for permission "to just hum" the national anthem.

6. *International Musician,* February 1944, 1; *Broadcasting,* January 15, 1937, 12; *Overture,* September 1950, 9. According to *Overture,* broadcasters owned 25 percent of the newspapers in the country. The number of newspaper-controlled stations is unknown, but it was large enough for Congress to investigate the matter in 1937. See also Sanjek, *American Popular Music,* 3:169.

7. *Broadcasting,* July 20, 1942, 55; *Overture,* September 1950, 9. Relations between Hubbard and the musicians' union in St. Paul were notoriously bad, and the source of much trouble for Petrillo.

8. *Broadcasting,* August 10, 1942, 9. Other labor leaders voiced views similar to Petrillo's. Walter Reuther, head of the United Automobile Workers, discussed technological change in "'Magic Carpet' Economics Are Not Good Enough," reprinted in *Machines and the Man: A Sourcebook on Automation,* ed. Robert P. Weeks (New York: Appleton-Century-Crofts, 1961), 202–17. For the views of James Carey, head of the International Union of Electrical Workers, see Carey, "Labor's View," in ibid., 96–102. Also David F. Noble, *America by Design: Science, Technology, and the Rise of Corporate Capitalism* (New York: Oxford University Press, 1977), xvii–xxvi.

9. Vern L. Countryman, "The Organized Musicians" (part 2), *University of Chicago Law Review* 16 (Winter 1948): 256; *International Musician,* December 1943, 1; Sanjek, *American Popular Music,* 3:205. As many as thirty-nine states passed legislation aimed at ASCAP practices.

10. "Memorandum of the United States in Opposition to Defendants' Motion to Dismiss the Complaint and in Support of the Government's Application for a Preliminary Injunction," in District Court of the United States for the North District of Illinois, Eastern Division (Civil Action No. 4541, Filed October 1, 1942), 1–2 (hereafter cited as District Court Papers). In July 1942 Attorney General Francis Biddle authorized Arnold to file a complaint against the AFM charging the union with violation of section 1 of the Sherman Act. NAB officers denied that they had initiated the legal action against the union, but the close contact between Arnold and NAB officials suggests otherwise. *Broadcasting* reported that the meeting with Arnold was his idea, noting that Arnold believed that a case against the AFM might be a means to overturn recent court rulings concerning "make-work" issues. NAB officials agreed to gather evidence for a possible suit. *Broadcasting,* August 17, 1942, 59. The Supreme Court had given tacit approval to standby orchestras on March 2, 1942, when it upheld the right of the International Brotherhood of Teamsters, Chauffeurs, Stablemen and Helpers to require the use of its members on trucks entering New York. *Broadcasting,* March 9, 1942, 1.

11. *International Musician,* August 1942, 1, 3. See also Leiter, *Musicians and Petrillo,*

149–53. Some of the musicians in the Interlochen band were not amateurs but older, accomplished musicians who taught at the camp.

12. *International Musician,* August 1942, 1.

13. *Broadcasting,* August 31, 1942, 1. The union argued that broadcasters had exaggerated the situation "beyond all reasonable proportions." *International Musician,* November 1942, 19.

14. *International Musician,* January 1944, 1. The text of the Senate resolution to probe Petrillo is in *Broadcasting,* August 31, 1942, 9, 11; and September 7, 1942, 7. Clark's bill was Senate Resolution 286.

15. *Broadcasting,* September 28, 1942, 55; "Use of Mechanical Reproduction of Music," *Hearings before a Subcommittee of the Committee on Interstate Commerce, United States Senate,* 77th Cong., 2d sess., 1942, 57–64 (hereafter cited as *Senate Hearings,* 1942).

16. *Broadcasting,* September 21, 1942, 11; *Senate Hearings,* 1942, 3–12; *International Musician,* August 1942, 1. Davis reminded the subcommittee that he had already asked Petrillo to rescind the ban but that Petrillo had refused.

17. *Broadcasting,* September 21, 1942, 63; *Senate Hearings,* 1942, 13–55. Fly accumulated his statistics after Senator Vandenberg, responding to the Interlochen dispute, requested that the FCC investigate the practices of Petrillo. Fly noted that 723 of the nation's 796 radio stations had responded to an FCC questionnaire that had been mailed to broadcasters. The survey indicated that network affiliates in smaller cities were more likely to rely on recorded music than those in larger cities. It also showed that more than half of all recorded music was from transcriptions rather than phonograph records. Many of the seventy-three stations not responding to the FCC questionnaire were no doubt stations without staff orchestras. The survey showed that 463 of the stations responding employed no full-time musicians, and 124 stations employed only one musician.

18. *Broadcasting,* September 28, 1942, 57. "If it could be measured," Padway said, the union's contribution to the war effort would amount to "hundreds of thousands of dollars." *Senate Hearings,* 1942, 110–11. Also Hitchcock interview, August 19, 1993. RCA-Victor, Columbia, Muzak, and World Broadcasting produced "V-disks" for the military. The first shipment went out in 1943. Roughly a third of the recordings were of new material; the remainder came from radio programs. V-disks could be distributed to U.S. hospitals that sheltered wounded soldiers. Hitchcock reported that musicians were eager to record V-disks, even though they did so without charge. "Musicians wanted to play," he said. Hitchcock himself recorded with the "V-Disks All Stars," a sixteen-piece band that included Wilbur "Willie" Schwartz, Ted Nash, and Mahlon Clark.

19. *Broadcasting,* September 28, 1942, 9, 57.

20. District Court Papers, 16–17. In stressing the importance of recorded music to the industry, Arnold provided a close look at the nature of the recording industry. His memorandum explained that the nation's record companies were licensed by the AFM to employ union members. Using the services of AFM members, the record companies made "master discs," from which millions of reproductions could be made. Subsidiary distributors sold these reproductions to retailers or "jobbers," who resold the records to retailers. Arnold noted that of the one hundred thousand records sold annually, 80 per-

cent went to retailers who sold the records to the general public for home use, 19 percent went to operators of jukeboxes, and 1 percent went to radio broadcasting companies.

21. *International Musician,* November 1942, 1, 19. Arnold suggested that 60 percent of AFM members worked outside music and thus were not unemployed. He noted that more than two hundred of the small stations that employed no musicians operated in areas where no musicians' union existed.

22. "Affidavits Filed by the United States in Support of Request for Preliminary Injunction," District Court Papers, 66–69. Miller estimated that 60 percent of the members of AFM locals were not professional instrumentalists but people who made their living largely in other occupations. According to Miller, Petrillo's tactic of withdrawing the services of name bands from the networks was an attempt to force businesses to conspire against one another, and to "deprive broadcasting stations of services vitally necessary to their continued existence." Miller told Judge Barnes that 176 radio stations broadcast in areas in which no musicians' union existed. In those areas, Miller said, broadcasters were "wholly dependent for their continued existence upon the use of recorded music."

23. Ibid., 95, 114. "Only about 10%" of the members in the local musicians' union, Le Poidevin suggested, were "professional musicians who rely on music for a livelihood." Le Poidevin said his station needed a "fresh supply" of recordings to survive. "Without these mechanical records," he said, "the quality of our programs will deteriorate."

24. Ibid., 142.

25. Ibid., 224–25. Pressley's business had cabins in back and two gas pumps in front. His clientele consisted of truck drivers and travelers. He said his dance floor was "entirely too small to accommodate a large enough crowd for an orchestra."

26. District Court Papers, 114.

27. *International Musician,* November 1942, 1; *Broadcasting,* October 19, 1942, 9, 46–48.

28. *Broadcasting,* October 19, 1942, 49; *International Musician,* November 1942, 19.

29. See Donald Reid, "Reflections on Labor History and Language," in *Rethinking Labor History: Essays on Discourse and Class Analysis,* ed. Lenard R. Berlanstein (Urbana: University of Illinois Press, 1993), 39–54.

30. *Official Proceedings,* 1944, 54–56; *Broadcasting,* January 25, 1943, 8. The Pittsburgh dispute led to a three-month agreement, during which the staff orchestras on both stations worked according to previous labor contracts.

31. *Broadcasting,* December 7, 1942, 9. Petrillo appeared before the Bingham committee on January 12, 1943.

32. Ibid., January 18, 1943, 38–42. Petrillo said that Arnold had threatened him with "five years in the can" if he participated in a secondary boycott. AFM counsel Padway suggested that 1 or 2 cents per record would be a reasonable royalty fee.

33. *Broadcasting,* December 7, 1942, 38–42; *International Musician,* July 1943, 1. Padway strengthened this argument by submitting a fifty-page statement explaining that "monopoly, interlocking arrangements, and large profits" characterized broadcasting and recording. This statement attacked the "handful of tremendously wealthy and powerful corporations" that allegedly dominated broadcasting and recording. Petrillo pointed out that 201 network affiliates were paying nothing to musicians.

34. *Broadcasting,* January 18, 1943, 9, 28, 39.

35. Ibid., October 26, 1942, 56; *Overture,* November 1942, 2. "It will do us no good to destroy Petrillo as an individual," Rosenbaum stated. "There will only rise up others to succeed him . . . who will be even more determined and unyielding." Refuting the position of the majority of broadcasters, Rosenbaum maintained that musicians had a "just complaint." Phonograph recordings, he admitted, "when used for profit, destroy the further employment of the very men who made them."

36. *Broadcasting,* October 26, 1942, 56, 58.

37. Ibid., February 1, 1943, 14. Annual profits of the jukebox industry were estimated to be $150 million.

38. Ibid., March 1, 1943, 9; March 22, 1943, 39–40. The AFM reminded the industry that the recording ban had the support of musicians' unions in England and in several nations in South America, all of whom agreed not to export recordings to the United States for the duration of the ban.

39. Ibid., April 19, 1943, 9; May 17, 1943, 16; May 24, 1943, 11. Petrillo initially spurned industry overtures, explaining that there was no reason to meet until record companies offered a constructive proposal; but he changed his mind. The union planned to tie the size of the required orchestras to the wattage of the radio stations. More powerful stations would employ larger orchestras. In a concession to the union, industry leaders agreed to conduct separate negotiations with manufacturers who produced popular records and with those who made long-playing transcriptions. (Dozens of small firms produced only one of these types of recordings.) The concession, however, proved of little value. On April 19, manufacturers of phonograph records used in homes, jukeboxes, and radio rejected the union's fixed-fee plan. A meeting with transcribers, held in early May, also proved fruitless.

40. *International Musician,* March 1943, 1; May 1943, 1; *Broadcasting,* May 3, 1943, 74. In late April Judge Barnes once again dismissed an antitrust suit that Arnold had filed.

41. *Broadcasting,* May 24, 1943, 11; *International Musician,* July 1944, 1. The WLB intervened in dozens of labor-management conflicts.

42. *International Musician,* October 1943, 1; January 1944, 1; *Official Proceedings,* 1946, 86. The WLB began hearings on the record dispute in July 1943. Only 5 percent of the fund could be used by the musicians' union for administrative purposes. Musicians received union wages for Record and Transcription Fund performances. According to the union, the fund offered partial compensation for technological displacement and promised to "foster musical culture."

43. Interview with Milton Gabler, Honolulu, July 13, 1994. Gabler, who became the vice president of the Artists and Repertoire Department, said that Jack Kapp, "a Chicago man," was "friendly with Petrillo."

44. *International Musician,* July 1944, 1; Gabler interview, July 13, 1994. Recording companies realized that music-starved consumers might spend millions on new recordings once the boycott ended. All of them hoped to benefit from the resulting spending spree.

45. *Overture,* July 1944, 11; *Official Proceedings,* 1945–46, 85. The WLB decision to order musicians back to work was reached by a vote of eight to four, with labor representatives on the board dissenting.

46. *Official Proceedings,* 1945–46, 79–82; *International Musician,* November 1944, 1. The WLB sent "the canned music controversy" to the Economic Stabilization Board, which passed the problem on to Roosevelt.

47. Gabler interview, July 13, 1994. Vinylite was apparently used in the production of most transcriptions, including V-disks. It too was difficult to procure during the war. Transcription companies often relied on reprocessed Vinylite to make new recordings. Sanjek, *American Popular Music,* 3:220. Also Leiter, *Musicians and Petrillo,* 137.

48. *Overture,* December 1944, 4; *Official Proceedings,* 1945–46, 86–87; 1946–47, 102; *International Musician,* February 1947, 5. Fees ranged from a quarter of a cent for each 35-cent record to 5 cents for each $2 record. For library transcription services, which rented transcriptions to radio, the fee amounted to 3 percent of gross revenues derived from the use of the material. To ensure the success of the plan, the industry agreed to give the union access to its financial books; this alone represented a major concession, and evidence of union power. Some musicians complained that the companies should pay higher fees, but Petrillo argued that the amount of the fees was far less important than the principle established.

49. *Official Proceedings,* 1945–46, 87. The fund proved more democratic in its effect than the earlier national radio agreements, which helped musicians only in cities where radio stations operated. The radio agreement was of little value for black musicians, since discriminatory employment practices excluded them from the newly generated radio jobs. Revenue from the unemployment fund, however, made its way to black as well as white locals. Indeed, questions about distributing the fund fairly apparently helped twelve "colored locals" under the control of white unions gain full autonomy. At the union's 1944 national convention the national executive board abolished subsidiary charters, thus enabling black locals to "govern themselves in the same manner as any other local." A delegation of African Americans had requested the change. The delegation consisted of R. L. Goodwin, P. S. Cooke, Harry W. Gray, L. V. Plummer, Edward Bailey, and William H. Bailey, all whom were officials of black unions in large cities. *Official Proceedings,* 1945–46, 69; *International Musician,* July 1944, 1.

50. *Official Proceedings,* 1945–46, 87–88. "Reactionary employers," he said, "have acted as though the musicians they wished to employ were their vassals and slaves." Petrillo also compared employers with nineteenth-century steelmakers, who "stopped at nothing to defeat the just demands of their employees." "Honesty and fairness," Petrillo concluded, "have triumphed over falsity and fraud."

51. *Broadcasting,* February 22, 1943, 11; December 7, 1942, 38–42. AFM counsel Padway had predicted that the ban would produce between twenty thousand and sixty thousand jobs.

Chapter Seven: Balancing Success and Failure

1. *Official Proceedings,* 1945–46, 48–53; 1946–47, 57–65; 1947–48, 115–16; *Variety,* January 28, 1948, 30; June 9, 1948, 39. Petrillo reported that twenty-five hundred musicians worked "under regular weekly salaries in radio stations for periods ranging from 30 to 52 weeks a year." *International Musician,* April 1944, 1. Also John Scott Kubach, "Unemployment and the American Federation of Musicians: A Case Study of the Eco-

nomic Ramifications of Technological Innovations and Concomitant Governmental Policies Relative to the Instrumental Employment Opportunities of the Organized Professional Musicians" (master's thesis, Ohio University, 1957), 208. Kubach reports that in 1945 there were 2,532 musicians employed fifty-two weeks a year in radio staff orchestras. An additional 5,976 musicians and 175 leaders also worked in single-engagement positions.

2. *Film Music Notes* 12 (March–April 1953): 23; *Overture*, May 1946, 8.

3. Tom Lewis, *Empire of the Air: The Men Who Made Radio* (New York: Harper-Collins, 1991), 303.

4. *Official Proceedings*, 1945–46, 48–49.

5. Ibid.

6. The following year, when motion-picture firms began investing in television, the union also prohibited film studios from using recorded music for television programming. *Overture*, February 1949, 5; *Official Proceedings*, 1945–46, 92; 1947–48, 98; Robert D. Leiter, *The Musicians and Petrillo* (New York: Bookman Associates, 1953), 172.

7. Interviews with I. B. "Buddy" Peterson, Will Brady, Roc Hilman, and Bob Fleming, Honolulu, October 8, 1993.

8. *Official Proceedings*, 1945–46, 48–50. "Beginning Monday, October 29," Petrillo's telegram stated, "wherever musicians play for FM broadcasting and AM broadcasting simultaneously, the same number of men must be employed for FM broadcasting as are employed for AM broadcasting, which means a double crew must be employed." The new union policy regarding the right of locals to negotiate with independent FM broadcasters reversed an earlier policy that gave the executive board control over contracts for FM broadcasting.

9. *New York Times*, October 25, 1945, 23. In 1945 ABC had no affiliates capable of receiving FM broadcasts. NBC and CBS were the primary beneficiaries of duplicating programs over AM and FM channels.

10. Federal Communications Commission, *Public Service Responsibility of Broadcast Licensees* (Washington, D.C.: FCC, March 7, 1946), 39; Richard J. Meyer, "The Blue Book," *Journal of Broadcasting* 6 (Summer 1962): 197–203.

11. Richard J. Meyer, "Reaction to the 'Blue Book,'" *Journal of Broadcasting* 6 (Fall 1962): 296–97.

12. United States House of Representatives, *House Reports*, 79th Cong., 2d sess., January 14–August 2, 1946, *Miscellaneous*, 1:9–10 (hereafter cited as *House Reports*, 79th Cong.). Also *Labor Law* (New York: Commerce Clearing House, 1972), 1719–20; *Billboard*, June 21, 1947, 36; *Overture*, August 1947, 6; and Leiter, *Musicians and Petrillo*, 158–59. The bill would also have made it illegal for radio workers to restrict "the production, sales, or use of records or transcriptions" or to demand "tribute" for "materials used or intended to be used in broadcasting." These provisions raised questions about musicians' ability to oppose the recording of radio shows or to demand royalties from such recordings. The congressional conference changed the word *tribute* to *exactions*.

13. *Congressional Record*, 79th Cong., 2d sess., February 21, 1946, 92, pt. 2:1543, 1550, 1553, 1556, 1558, 1564.

14. *House Reports*, 79th Cong., 5; *Congressional Record*, February 21, 1946, 92, pt. 2:1558.

15. *Congressional Record*, February 21, 1946, 92, pt. 2:1547, 1554, 1559, 1563; *Congres-*

sional Directory, 79th Cong., 2d sess., 1946, 149–56. Cellar represented Brooklyn, New York.

16. *Congressional Record,* 79th Cong., 2d sess., February 27, 1946, 92:1709–10; March 12, 1946, 92:2173–74; March 29, 1946, 92:2821–23; April 6, 1946, 92:3241–58; April 17, 1946, 92:3829. In a last-gasp effort to block the Lea bill, Marcantonio called it "the most extreme antilabor legislation that has ever been passed." Truman, who generally supported labor—and even played the piano—undoubtedly realized that a presidential veto would be overturned.

17. *Overture,* July 1947, 15; *New York Times,* June 4, 1946, 18.

18. *Broadcasting,* November 8, 1948, 22, 59. This radio trade journal discussed political figures reelected to Congress. In the House, Representative Capper owned WIBW in Topeka, Kansas; Ellsworth owned KRNR in Roseburg, Oregon; O'Konski owned WLIN in Merrill, Wisconsin; and Phillips owned KPAS in Benning, California. Senator Gurney held interests in WNAX in Yankton, South Dakota; and Senator Knowland was part owner of KLX in Oakland, California.

19. Ibid.; April 5, 1948, 34; April 19, 1948, 23; June 7, 1948, 4; December 13, 1948, 26. Before entering politics, Representative Karl Stefan of Nebraska had been a radio news commentator, and Senators Glen H. Taylor of Idaho and W. Lee ("Pass the Biscuits, Pappy") O'Daniel of Texas had worked as radio entertainers. Vandenberg's friend owned stations WKZO and WJEF. Hulbert Taft, Jr., owned station WKRC in Cincinnati and its FM outlet, WCTS. David Taft managed WCTS. Also *Billboard,* June 21, 1947, 2; and *International Musician,* September 1947, 20.

20. *House Reports,* 79th Cong., 3.

21. *Overture,* August 1947, 6.

22. *Chicago Tribune,* May 29, 1946, 1; *New York Times,* June 4, 1946, 18.

23. *Chicago Tribune,* May 30, 1946, 9.

24. *Billboard,* June 21, 1947, 36; *Labor Law,* 1801. Citing a 1928 ruling involving a carpenters' union, the judge emphasized, "Every man has a full freedom in the disposal of his labor according to his will, and workmen have a right to organize for the purpose of promoting their common welfare by lawful means. They may impose any condition of their employment which they may regard as beneficial to them." La Buy ruled that the Lea Act violated the First Amendment protection of freedom of speech by impinging upon the musicians' right to picket. See also *Chicago Tribune,* December 3, 1946, 24.

25. Petrillo must have wondered whether the new chief justice, Fred M. Vinson, would support the union's position. In 1943, when Vinson headed the War Labor Board, he had questioned the loyalty of Petrillo and demanded that he call off the recording ban. Petrillo's response to Vinson was unflattering. Vinson voted with the Court majority in 1947, as had Justices Black, Frankfurter, Burton, and Jackson. Justice Douglas did not participate in the case. The dissent of Justices Reed, Murphy, and Rutledge argued that the Lea Act was unconstitutional. It was unreasonable to expect "man or jury," the dissent read, to know how many employees are needed to perform services in broadcasting. The minority maintained that questions concerning the desired quality of programs, the skill levels of employees, and the number of hours that employees worked seriously complicated the ability to determine desired levels of employment. See *Overture,* August, 1947, 6; and *Official Proceedings,* 1946–47, 92–93.

26. *Congressional Record,* 80th Cong., 1st sess., 1947, 63, pt. 5:6446. Unlike the Lea Act, Taft-Hartley authorized no criminal penalties for demands that employers hire "unneeded" employees. Section 8(b) prevented labor organizations from attempting to force employers to "pay or deliver any money or other thing of value, in the nature of an exaction, for services which are not performed." Also "Labor Management Relations Act, 1947," Public Law 101, 80th Cong., 1st sess., 24, Papers of Local 47. Also *Labor Law,* 6121; *Official Proceedings,* 1948, 37–40; *International Musician,* February 1948, 17; *Overture,* July 1947, 15; August 1947, 6; and *Billboard,* June 21, 1947, 22.

27. "Report on Stations," by Phil Fischer, Papers of Local 47. On union membership, see *Overture,* August 1948, 18; Everett Lee Refior, "The American Federation of Musicians: Organization, Policies, and Practices" (master's thesis, University of Chicago, 1955), 55; *Official Proceedings,* 1944, 7–27; and 1948, 7–32.

28. *New York Times,* June 4, 1946, 1, 18; *Chicago Tribune,* June 5, 1946, 6.

29. "Minutes of Special Meetings of the International Executive Board," *International Musician,* January 1947, 5, 9, 38; *Variety,* November 26, 1947, 23, 42; December 10, 1947, 32.

30. "Minutes of Special Meetings," 40. The motion gave the executive board power to tell AFM members "to refrain from rendering services for any or all types of recording." The board also received permission to enter "into the music-recording business in direct competition with other recording companies if, in the wisdom of the executive board, such action should be necessary to protect the interests of members of the AFM." Union officers no doubt realized that such action was likely to result in antitrust proceedings.

31. *Variety,* December 3, 1947, 27; December 31, 1947, 21.

32. Ibid., December 24, 1947, 25, 32. Verne Burnett headed the public relations subcommittee. Burnett had formerly directed radio advertising for General Foods and worked as a public relations expert for General Motors and the copper industry. The organization selected as legal counsel Sydney Kaye, who worked for Broadcast Music Incorporated.

33. Russell Sanjek, *American Popular Music and Its Business: The First Four Hundred Years* (New York: Oxford University Press, 1988), 3:229. Sanjek notes that Decca was best prepared for a strike, since it had a large collection of unreleased recordings.

34. Correspondence of Phil Fischer, 1, Papers of Local 47.

35. Ibid.; *Overture,* April 1948, 6; *Variety,* November 26, 1947, 23, 42. Niles Trammell and Frank Mullen represented NBC, Frank White CBS, Mark Woods ABC (the former vice president and treasurer of NBC became president of ABC with the incorporation of the company in 1942), and Robert Sweezey Mutual. Theodore Streibert of station WOR in New York also attended. Also *Broadcasting,* November 24, 1947, 13.

36. *Broadcasting,* November 24, 1947, 13; *Overture,* April 1948, 6. An additional request made by the networks concerned their efforts to adopt the "Chicago formula" for hiring staff orchestras in Los Angeles and New York. In Chicago, where 180 musicians worked full-time in staff orchestras yet very few musicians worked in commercial programs, the networks had the right to purchase the services of staff musicians and sell the services for any price to sponsors. The policy often produced profits for the networks. In Los Angeles, where ninety musicians worked full-time in staff orchestras and five hundred to six hundred musicians played on commercial programs, union

rules more clearly separated the roles of full-time and part-time musicians. Los Angeles forbade the networks to resell the services of full-time "sustaining" staff musicians, a policy designed to maximize employment for staff musicians. In New York 260 musicians worked in staff orchestras. That city had fewer commercial programs than Los Angeles and allowed stations to sell staff musicians for commercial programs, as long as this did not reduce employment opportunities for staff musicians.

37. *Variety*, November 26, 1947, 23. The networks also asked to use musicians for both "cooperative" and "participating" programs. For years the union had prohibited musicians from working on both types of programs. Its opposition to cooperative shows—network programs sponsored by several local companies on a regional basis—arose from concern for the jobs of instrumentalists in smaller stations. In contrast, opposition to participating shows—network programs sponsored by two or more national advertisers—stemmed from concern for big-city musicians, who stood to lose job opportunities if two advertisers already sponsoring separate programs decided to share ownership of a single program. Also *Broadcasting*, November 24, 1947, 13; and *Overture*, April 1948, 6. Representatives suggested that television would attract larger audiences if broadcasters could televise AM radio programs.

38. Correspondence of Phil Fischer, 3, Papers of Local 47. The remanding of the WAAF case to Judge La Buy's Chicago court had forced Petrillo to schedule new labor negotiations in his home town. *Overture*, April 1948, 6; *Variety*, November 26, 1947, 42. In an apparent effort to open the door to compromise, Petrillo made a single, significant concession to broadcasters. Since the networks were already carrying several cooperative programs featuring all-vocal groups not covered by AFM restrictions, Petrillo agreed to lift restrictions on cooperative broadcasting. The move had the effect of reducing job opportunities for musicians in smaller cities, but it served to increase radio employment in major media centers. Local 47 correctly predicted that cooperative programming would produce additional employment for approximately fifty musicians in Los Angeles. The press viewed the AFM's reversal on cooperative programming as a minor retreat, but Petrillo had long championed the ambitions of instrumentalists in less populated areas. He had launched the long wartime recording ban largely for their benefit. Sacrificing a measure of control over smaller job markets suggested that the AFM had become less capable of shaping labor relations.

39. *Variety*, November 26, 1947, 42.

40. *Variety*, December 10, 1947, 29, 32. "I think that Mr. Petrillo," Fischer explained, "was anxious to know the outcome of the case against him before proceeding any further. I think he felt that if the outcome of the trial was favorable . . . the Federation's position would be much stronger." Correspondence of Phil Fischer, 4, Papers of Local 47. Also *Official Proceedings*, 1947–48, 92–93; and House Committee on Education and Labor, *Restrictive Union Practices of the American Federation of Musicians*, 80th Cong., 2d sess., 1948, 381 (hereafter cited as *Restrictive Union Practices*). WAAF had apparently made $180,000 in profits after taxes without live music.

41. "Report on Stations," Correspondence of Phil Fischer, 2, Papers of Local 47.

42. United States House of Representatives, "Investigation of James C. Petrillo and the American Federation of Musicians," *House Reports*, 80th Cong., 1st sess., January 3–December 19, 1947, *Miscellaneous*, 6:11–13 (hereafter cited as *House Reports*, 80th Cong.). See also *Official Proceedings*, 1947–48, 94; and *Variety*, December 17, 1947, 43.

The Kearns report suggested that before strikes, the National Labor Relations Board should conduct investigations to decide whether the majority of employees approved the strike. The Kearns committee held meetings in Los Angeles and New York as well as Washington, D.C.

43. *Restrictive Union Practices*, 3–5, 29. An executive from the Radio Manufacturers Association appeared, and the more recently formed Television Broadcasters Association and FM Association also sent their presidents to appear before the committee. Witnesses generally agreed that the practices of the AFM had hindered the progress of industry and the interests of consumers. As James Murray, vice president of RCA, said, "It is difficult to see how Mr. Petrillo is doing anyone any good."

44. *Official Proceedings*, 1947–48, 93–94; *Washington Post*, January 22, 1948, 11. The *Post* noted that Petrillo testified "under a battery of klieg lights and before a battery of microphones." Photos in *Broadcasting*, January 26, 1948, 14, verify these statements. Also *Restrictive Union Practices*, 355, 377.

45. *Official Proceedings*, 1947–48, 94; *Restrictive Union Practices*, 355, 337, 392, 396–97, 399; *Washington Post*, January 22, 1948, 11, 15B.

46. *Variety*, January 28, 1948, 30; *Washington Post*, January 22, 1948, 11; *Broadcasting*, January 26, 1948, 14, 82.

47. *Restrictive Union Practices*, 403–4.

48. "Statement of Milton Diamond to the Committee on Education and Labor of the House of Representatives," 3–12, Papers of Local 47. Diamond presented figures detailing the distribution of income within entertainment industries. He pointed out that members of radio staff orchestras now received only 3.5 percent of the radio industry's gross revenues, and musicians "other than leaders" secured but 1 percent of total record industry revenues. Diamond also noted that jukebox companies now collected over $230 million annually from four hundred thousand coin-operated machines, yet the companies "do not pay one penny to musicians." Although total sales of record companies in 1946 topped $165 million, the entire record industry paid musicians, excluding bandleaders, only $1.6 million. In regard to the motion-picture industry, Diamond explained that the salaries MGM and Columbia Pictures paid to musicians equaled 0.9 percent of the total expenditures of each firm. Before taxes, the profits of MGM studios were eighteen times the amount paid to musicians. Columbia's profits before taxes were 11.3 times its expenditure for musicians. Diamond stated that the film industry offered casual employment to 5,518 musicians, but "only 239 received full time employment." He asked the committee to consider "these cold facts" before passing judgment.

49. *House Reports*, 80th Cong., 4. Diamond defended the union as a democratic institution, explaining that musicians had a greater voice in the affairs of their union than did the stockholders of major corporations.

50. Howell John Harris, *The Right to Manage: Industrial Relations Policies of American Business in the 1940s* (Madison: University of Wisconsin Press, 1982), 186–87.

51. *Variety*, January 28, 1948, 27, 30; February 4, 1948, 26, 38.

52. Ibid., February 11, 1948, 23; March 3, 1948, 24.

53. Ibid., March 24, 1948, 27, 33. Also *Overture*, April 1948, 6–7. The final bargaining sessions did not end until March 17, eleven days after they began. The union also lifted certain restrictions on participating programs.

54. *Variety,* March 3, 1948, 38; interview with Lenny Atkins, Honolulu, October 8, 1993.

55. Fleming, Hilman, and Brady interviews, October 8, 1993.

56. *Official Proceedings,* 1948–49, 103–4. The companies were Associated Program Service (Muzak), Lang-Worth Feature Programs, and Standard Radio Transcription Services. The first two companies filed charges against the AFM and New York Local 802. Standard, which had for some time been a thorn in the side of the union, filed charges in Los Angeles against the AFM and Local 47. On May 27 AFM lawyers countered accusations of the three transcription companies by submitting to the NLRB a legal defense of the union's position. Union arguments maintained that the recording ban was a direct dispute between musicians and record companies, and not a secondary boycott. The labor board's investigation continued for the next several months, requiring several meetings among AFM attorneys, NLRB officials, and legal representatives of the transcription companies.

57. *Variety,* December 17, 1947, 43; April 28, 1943, 46. Even Representative Kearns, in his harsh subcommittee report of 1947, had suggested changing the 1909 law in order to help musicians "make up" for revenue lost as a result of the Taft-Hartley Act. The jukebox industry, he stated, "has made great profits by capitalizing on the playing of recordings made by AFM members." Although Congress had discussed similar proposals for years, AFM officers believed that now, having recently outlawed the record-royalty fund, Congress would finally amend the old copyright law.

58. Sanjek, *American Popular Music,* 3:433–35; *Variety,* May 19, 1948, 43.

59. *Variety,* June 9, 1948, 39, 44. The $26 million figure contradicts union reports in *Official Proceedings,* 1948–49, 122–23, which suggest that total earnings in radio for 1947 were closer to $22 million.

60. *International Musician,* January 1949, 6–8; *Official Proceedings,* 1947–48, 103–4. On July 21, 1948, Associated Program Service (Muzak) withdrew its charges against the AFM. Charles T. Douds was the regional director of the NLRB. On December 10 Solicitor of Labor William S. Tyson informed Secretary of Labor Maurice J. Tobin that the fixed-fee proposal did not conflict with the Taft-Hartley Act. On December 13 Attorney General Tom C. Clark agreed.

61. *Overture,* July 1949, 15; *International Musician,* February 1949, 34–35; March 1949, 7; October 1955, 8–9. Future trustees were to be appointed by the secretary of labor. Initially the Taft-Hartley Act stipulated that union unemployment funds could be used to help only employees of the employers who contributed to the fund. In other words, only musicians who made records could benefit from the AFM's unemployment fund. This was clearly contrary to the original intent of the fund, which was meant to aid instrumentalists harmed by the use of recorded music. The Supreme Court, however, interpreted the new law in such a way that all musicians could benefit from the unemployment fund. *National Labor Relations Board v. E. C. Atkins and Co.,* 331 U.S. 398, 403, May 19, 1947. The new trust fund could be used to help nonunion as well as union musicians, but the union's grip on professional musical services and the logistical problems associated with distributing funds to nonunion musicians ensured that AFM members would be the primary beneficiaries of the new MPTF.

62. *Official Proceedings,* 1947–48, 114–16; 1948–49, 122–23; 1949–50, 108–9; 1950–51, 128–29; 1951–52, 144–45.

Conclusion

1. John Scott Kubach, "Unemployment and the American Federation of Musicians: A Case Study of the Economic Ramifications of Technological Innovations and Concomitant Governmental Policies Relative to the Instrumental Employment Opportunities of the Organized Professional Musicians" (master's thesis, Ohio University, 1957), 220. Kubach estimates that the number of musicians who earned the majority of their income from musical performances declined by 51 percent from 1930 to 1954.

2. "Report of Records and Transcriptions Committee," poster in Bagley Collection.

3. *International Musician,* October 1951, 9.

4. Interview with Lenny Atkins, Honolulu, October 8, 1993. The record industry has managed to reduce the size of its contributions to the MPTF since 1950 and the future of the fund is uncertain.

5. *Broadcasting,* January 18, 1943, 38–43.

6. *Official Proceedings,* 1938–39, 80.

7. Howell John Harris, *The Right to Manage: Industrial Relations Policies of American Business in the 1940s* (Madison: University of Wisconsin Press, 1982), 94–95, 178–79.

8. Interview with Phil Fischer, Los Angeles, April 3, 1987.

9. *Official Proceedings,* 1958–59, 96, 147.

10. Ibid., 1938–39, 71–72; *Broadcasting,* January 18, 1943, 38–43. At midcentury the AFM demanded that Congress "face up to its responsibility to preserve music and the arts. "Diminuendo," 23, pamphlet prepared for the AFM by Hal Leyshon and Associates and printed by the International Press, Newark, N.J., Papers of Local 10, Chicago.

11. National Association of Manufacturers, "Calling All Jobs: An Introduction to the Automatic Machine," pamphlet (October 1957), reprinted in Robert P. Weeks, ed., *Machines and the Man: A Sourcebook on Automation* (New York: Appleton-Century-Crofts, 1961), 171–76.

12. David Landes, *The Unbound Prometheus: Technological Change and Industrial Development in Western Europe from 1750 to the Present* (New York: Cambridge University Press, 1969), 7.

Essay on Sources

THIS BIBLIOGRAPHIC ESSAY does not identify all sources consulted in the preparation of this book. It does, however, point readers to major primary sources and to the most significant recent secondary literature that provided factual information or influenced my interpretations during preparation. For fuller reference to the historiography, readers should look at the extensive notes to the text chapters.

ARCHIVAL SOURCES on the history of musicians and their unions are scarce. The Charles Leland Bagley Collection at the University of Southern California's Regional History Center contains various records and newspaper clippings relevant to the rise of the National League of Musicians and its successor, the American Federation of Musicians (AFM). The collection contains copies of printed union constitutions, price lists, and bylaws as well as union financial records and convention proceedings to the 1960s. Researchers interested in AFM history should consult local branches of the union, a few of which have saved their old records and trade papers. For this project I found useful material at locals in Boston, Chicago, and Los Angeles. Various records of the local in Columbus are in the Archives–Library Division of the Ohio Historical Society in Columbus. Another useful source is the American Federation of Labor Records, The Samuel Gompers Era, available on microfilm at the University of California, Los Angeles, Special Collections Library. These records include correspondence of musicians, employers, and union leaders. One of the best sources on the history of the AFM is the federation's monthly newspaper, *International Musician,* available in several university and public libraries. The paper carries minutes of executive board meetings and convention proceedings as well as other union news.

Researchers should supplement union records and newspapers with industry trade journals. *Broadcasting, Billboard,* and *Variety* present in-depth looks at industrial developments and offer valuable insights into management's view of labor relations and government-business relations. They also reveal managerial strategies for continued economic growth. More obscure journals such as *Exhibitor's Herald, Film Daily,* and *Film Music Notes* are available at the University of Southern California's Cinema-Television Library and at the Academy of Motion Picture Arts and Sciences, both in Los Angeles. The Theatre Historical Society of America in Elmhurst, Illinois, holds copies of *Marque* as well as other material relevant to early film history. City newspapers and the labor press helped me verify information gleaned from all of these sources.

Like a growing number of labor studies, this project relied on oral history. I learned the value of this important investigative tool when Local 47 put me in touch with Phil Fischer and John TeGroen, two former union officials who were always generous with their time and energy. Fischer and TeGroen explained more clearly how industrial change altered musicians' lives and how the AFM responded to new business conditions. They also pointed me toward instrumentalists who generously shared their own career experiences. I contacted additional interviewees through other AFM locals and by matching names in union papers and records to those in current telephone directories. Tapes and notes from these interviews are presently in my possession, but I intend eventually to place them in a university library.

I also relied on government documents. Occupational statistics of the U.S. Census Bureau helped me estimate the number of professional musicians in America, and reports of the U.S. Department of Labor, published in *Monthly Labor Review*, documented the declining job opportunities in theaters during the 1920s and 1930s. Public documents were indispensable to reconstructing the history of musicians in the 1940s. The *Congressional Record* shed new light on industrial relations in the postwar years and showed that the AFM was a primary target in congressional efforts to roll back the power of organized labor. Published reports of hearings before the Interstate Commerce Commission and the House Committee on Education and Labor were equally informative, as were reports of the Federal Communications Commission.

LIKE PRIMARY SOURCES, the secondary literature on musicians and their unions is skimpy. John R. Commons, remembered for his institutional approach to the study of labor history, was one of the first scholars to study instrumental musicians as a labor group. "The Musicians of St. Louis and New York," *Quarterly Journal of Economics,* May 1906, makes it clear that despite their hardships, the problems of professional musicians were minor compared with those facing other workers in industrializing America. Vern L. Countryman's two-part essay, "The Organized Musicians," published in the *University of Chicago Law Review,* Autumn 1948 and Winter 1948, looks closely at the origins, structure, and early problems of the AFM. Coping with technological change, Countryman notes, had become the union's greatest challenge. The standard work on the AFM has been Robert D. Leiter, *The Musicians and Petrillo* (New York: Bookman Associates, 1953). Most important, Leiter's work provides a general outline of industrial relations during the war and postwar years. Though Leiter recognizes that technological change transformed the musicians' working world, his interest in the manifold actions of the union obscures the dramatic impact of that change as well as the musicians' efforts to cope with it.

Several unpublished works of the 1950s and 1960s have made additional contributions to the story of musicians and technological change. Abram Loft, "Musicians, Guild, and Union: A Consideration of the Evolution of Protective Organization among Musicians" (Ph.D. diss., Columbia University, 1950) places the early organizational efforts of American musicians within the context of European experiences. Everett Lee Refior's "The American Federation of Musicians: Organization, Policies, and Practices" (master's thesis, University of Chicago, 1955) discusses the nature of the musical workforce as well as the history of the union. John Scott Kubach's "Unemployment and the American Federation of Musicians: A Case Study of the Economic Ramifications of Technological Innovations and Concomitant Governmental Policies Relative to the Instrumental Employment Opportunities of the Organized Professional Musicians" (master's thesis, Ohio University, 1957) demonstrates quite conclusively that technological change rather than economic cyclical variations accounted for musicians' employment problems during the Great Depression. Robert Lee Humes's "Labor Relations and the American Federation of Musicians: Six Locals in Pennsylvania" (master's thesis, Pennsylvania State University, 1965) handles union activity at the local level.

More recent works flesh out musicians' work experiences. Robert R. Faulkner's *Hollywood Studio Musicians: Their Work and Careers in the Recording Industry* (Chicago: Aldine and Atherton, 1971) discusses working conditions and patterns of hiring in motion-picture studios, illuminating the daily challenges musicians faced in their efforts to establish recording careers. Sandy R. Mazzola's "When Music Is Labor: Chicago Bands and Orchestras and the Origins of the Chicago Federation of Musicians, 1880–1902" (Ph.D. diss., Northern Illinois University, 1985) portrays musicians at work in numerous other environments. In the process, Mazzola reveals basic differences between the labor of musicians and that of other groups of workers. I also benefited from two works by Neil Leonard: *Jazz: Myth and Religion* (New York: Oxford University Press, 1987), which describes the speech, dress, and habits of one group of musicians; and *Jazz and the White Americans: The Acceptance of a New Art Form* (Chicago: University of Chicago Press, 1962), which discusses the impact of mechanization on job opportunities and musical education. *Edison, Musicians, and the Phonograph: A Century in Retrospect,* ed. John Harvith and Susan Edwards Harvith (New York: Greenwood, 1987), a collection of interviews with musicians who worked in the early recording industry, also shows how technological change affected musicians' work habits. I gained additional insights from George T. Simon's *The Big Bands,* 4th ed. (New York: Schirmer Books, 1981), and Clifford McCarty, ed., *Film Music I* (New York: Garland, 1989).

Few published works have explored matters of race, ethnicity, and gender in musical life. Donald Spivey, *Union and the Black Musician: The Narrative of William Everett Samuels and Chicago Local 208* (New York: University Press of America, 1984), documents the life of an African American and AFM official in Chicago during the period under study. William Barlow, *Looking up at Down: The Emergence of Blues Culture* (Philadelphia: Temple University Press, 1989), and Susan Curtis, *Dancing to a Black Man's Tune: A Life of Scott Joplin* (Columbia: University of Missouri Press, 1994), also explore the working lives of black musicians. Steven Loza's *Barrio Rhythm: Mexican American Music in Los Angeles* (Urbana: University of Illinois Press, 1993), reveals the world of Mexican American musicians. Readers interested in the role of women in nineteenth-century music should see Michael Broyles, *"Music of the Highest Class": Elitism and Populism in Antebellum Boston* (New Haven: Yale University Press, 1992), and Craig H. Roell, *The Piano in America, 1890–1940* (Chapel Hill: University of North Carolina Press, 1989). Roell's work helps explain how social expectations shaped women's musical training.

THE LITERATURE on the film, radio, and recording industries is extensive. One of the more concise histories of the film industry is John Izod, *Hollywood and the Box Office, 1895–1986* (New York: Macmillan, 1988). Tino Balio, ed., *The American Film Industry* (Madison: University of Wisconsin Press, 1976), and Robert Sklar's *Movie-Made America: A Cultural History of the Movies* (New York: Chappell and Company, 1978) proved useful in the early stages of my research. Equally important was Douglas Gomery's work, particularly *Shared Pleasures: A History of Movie Presentation in the United States*

(Madison: University of Wisconsin Press, 1992), and "The Coming of Sound to the American Cinema: Transformation of an Industry" (Ph.D. diss., University of Wisconsin–Madison, 1975). Other important studies include Q. David Bowers, *Nickelodeon Theatres and Their Music* (Vestal, N.Y.: Vestal, 1986); Neal Gabler, *An Empire of Their Own: How the Jews Invented Hollywood* (New York: Anchor Books, 1988); and David Bordwell, Janet Staiger, and Kristin Thompson, *The Classical Hollywood Cinema: Film Style and Mode of Production to 1960* (New York: Columbia University Press, 1985). On the film industry in the postwar years, see Garth Jowett, *Film: The Democratic Art* (Boston: Little, Brown and Company, 1976).

Among the best works on the business and technological side of radio are Susan Smulyan, *Selling Radio: The Commercialization of American Broadcasting, 1920–1934* (Washington, D.C.: Smithsonian Institution Press, 1994); Susan J. Douglas, *Inventing American Broadcasting, 1899–1922* (Baltimore: Johns Hopkins University Press, 1987); Hugh G. J. Aitken, *The Continuous Wave: Technology and American Radio, 1900–1932* (Princeton: Princeton University Press, 1985); and Aitken, *Syntony and Spark: The Origins of Radio* (Princeton: Princeton University Press, 1976). Erik Barnouw's three-volume *A History of Broadcasting in the United States* (New York: Oxford University Press, 1966–70) is also useful. Michele Hilmes, *Hollywood and Broadcasting: From Radio to Cable* (Urbana: University of Illinois Press, 1990), and Philip Rosen, *The Modern Stentors: Radio Broadcasters and the Federal Government, 1920–1934* (Westport, Conn.: Greenwood, 1980), provide insights into the relationship between radio and the federal government. A newer work on this subject is Robert McChesney, *Telecommunications, Mass Media, and Democracy: The Battle for the Control of U.S. Broadcasting, 1928–1935* (New York: Oxford University Press, 1993). On radio programming, readers should see Smulyan, *Selling Radio* (cited above), and J. Fred MacDonald, *Don't Touch That Dial! Radio Programming in American Life, 1920–1960* (Chicago: Nelson-Hall, 1979).

Historians have shown less interest in the rise of the recording industry. For factual information I often relied on Russell Sanjek's three-volume *American Popular Music and Its Business: The First Four Hundred Years* (New York: Oxford University Press, 1988). Sanjek was one of the original employees of Broadcast Music Incorporated (BMI) and eventually served as the company's vice-president in charge of public relations. I also looked to Philip K. Eberly, *Music in the Air: America's Changing Tastes in Popular Music, 1920–1980* (New York: Hastings House, 1982), which traces the rise of the recording industry as well as the evolution of American music. Biographies of famous bandleaders and musicians also proved useful, including Robert Dupuis, *Bunny Berigan: Elusive Legend of Jazz* (Baton Rouge: Louisiana State University Press, 1993), and James Lincoln Collier, *Benny Goodman and the Swing Era* (New York: Oxford University Press, 1989). Readers interested in the origins of recording technology should see biographies of Thomas Edison. Martin V. Melosi, *Thomas A. Edison and the Modernization of America* (Glenview, Ill.: Scott, Foresman, 1990) is a good place to start. On the origins of magnetic recording, see William Charles Lafferty, Jr., "The Early Development of Magnetic Sound Recording in Broadcasting and Motion Pictures, 1928–1950" (Ph.D. diss., Northwestern University, 1981).

A DIFFERENT BODY of literature places the history of musicians within a broader perspective. The relationship between work and technological change has long been the focus of scholarly study. Karl Marx, *Capital,* vol. 1, published in German in 1867, maintained that capital's relentless use of labor-saving machinery tended to displace and demoralize workers. Harry Braverman's *Labor and Monopoly Capital: The Degradation of Work in the Twentieth Century* (New York: Monthly Review Press, 1974) renewed interest in this subject. Braverman emphasizes that management deploys new technology to separate the "conception" of work from its actual "execution." In *Contested Terrain: The Transformation of the Workplace in the Twentieth Century* (New York: Basic Books, 1979), Richard Edwards argues that the rise of impersonal bureaucracies as well as labor-saving machinery has increased management's control over the workforce. Michael Burawoy, in *The Politics of Production* (New York: Verso, 1985), and David Noble, *Forces of Production: A Social History of Industrial Automation* (New York: Oxford University Press, 1984), suggest that the total number of new skilled jobs technical innovation creates falls far short of the number it destroys. Several valuable articles on work and technological change appear in *Technology and Culture* 29 (October 1988).

The question of labor's response to technological change also has a long history. A path-breaking work in this area is David Brody's *Steelworkers in America: The Non-Union Era* (New York: Harper and Row, 1960), which shows how workers have used their limited power to improve working conditions. Equally influential is David Montgomery, *Workers' Control in America* (New York: Cambridge University Press, 1979). Montgomery focuses on shop-floor struggles to illustrate the various ways workers have been able to shape the production process. Other influential works that speak to the issue of labor response include Ronald W. Schatz, *The Electrical Workers: A History of General Electric and Westinghouse, 1923–60* (Urbana: University of Illinois Press, 1983); Sean Wilentz, *Chants Democratic: New York City and the Rise of the American Working Class, 1788–1850* (New York: Oxford University Press, 1984); and Steven J. Ross, *Workers on the Edge: Work, Leisure, and Politics in Industrializing Cincinnati, 1788–1890* (New York: Columbia University Press, 1985). Patricia A. Cooper, *Once a Cigar Maker: Men, Women, and Work Culture in American Cigar Factories, 1900–1919* (Urbana: University of Illinois Press, 1987), and Stephen H. Norwood, *Labor's Flaming Youth: Telephone Operators and Worker Militancy, 1878–1923* (Urbana: University of Illinois Press, 1990), are more sensitive to the subject of gender and labor resistance. Recent biographies have suggested that labor leaders generally accepted innovation as inevitable but tried to supervise reorganization of the workplace to protect as many jobs as possible. For examples, see Melvyn Dubofsky and Warren Van Tine, eds., *Labor Leaders in America* (Urbana: University of Illinois Press, 1987).

The literature on the role of the state in labor-capital relations is as controversial as it is extensive. Among the better historiographic works on this subject are Alan Dawley, "Workers, Capital, and the State in the Twentieth Century," in *Perspectives on American Labor History: The Problems of Synthesis,* ed. J. Carroll Moody and Alice Kessler-Harris (DeKalb: Northern Illinois University Press, 1990), 152–200; Thomas K. McCraw, "Regulation in America: A Review Article," *Business History Review* 49 (Summer 1975): 159–83; and Louis Galambos, "Technology, Political Economy, and Professionalization: Central Themes of the Organizational Synthesis," *Business History Review* 57 (Winter 1983): 471–93. Howell John Harris, *The Right to Manage: Industrial Relations Policies of American Business in the 1940s* (Madison: University of Wisconsin Press, 1982), proved particularly valuable for this project. Harris provides insights into the ideology of business leaders and public officials intent on rolling back the power of organized labor in the 1940s. An important new work on the subject of the state and industrial relations is Melvyn Dubofsky's *The State and Labor in Modern America* (Chapel Hill: University of North Carolina Press, 1994).

Finally, I have tried to place the history of musicians within the context of the sweeping cultural changes of the late nineteenth and early twentieth centuries. Many works have helped with this task, including Lary May, *Screening out the Past: The Birth of Mass Culture and the Motion Picture Industry* (Chicago: University of Chicago Press, 1983); Robert C. Allen, *Horrible Prettiness: Burlesque and American Culture* (Chapel Hill: University of North Carolina Press, 1991); Lizabeth Cohen, *Making a New Deal: Industrial Workers in Chicago, 1919–1939* (Cambridge: Cambridge University Press, 1990); Lawrence W. Levine, *Highbrow/Lowbrow: The Emergence of Cultural Hierarchy in America* (Cambridge: Harvard University Press, 1988); and Lewis A. Erenberg, *Steppin' Out: New York Nightlife and the Transformation of American Culture, 1890–1930* (Chicago: University of Chicago Press, 1981). Kathy Peiss, *Cheap Amusements: Working Women and Leisure in Turn-of-the-Century New York* (Philadelphia: Temple University Press, 1986); Roy Rosenzweig, *Eight Hours for What We Will: Workers and Leisure in an Industrial City, 1870–1920* (Cambridge: Cambridge University Press, 1983); and Gunther Barth, *City People: The Rise of Modern City Culture in Nineteenth-Century America* (New York: Oxford University Press, 1980), were also useful. So too was John F. Kasson, *Amusing the Million: Coney Island at the Turn of the Century* (New York: Hill and Wang, 1978).

Index

radio, 73, 104, 107, 144, 164, 188, **198, 199,** 235–36n. 1; in recording, 102, 164; in theaters, 33, 37–39, 88

Ethnicity: and studios, 100, 101–2; and unionization, 18–19, 208n. 22, 210n. 42

Federal Communications Commission (FCC): and Blue Book, 168; on musical employment, 144–45, 232n. 17; and National Plan, 125; origins of, 77; reverses policy on AM-FM duplication, 165

Federal Radio Commission (FRC), 77, 219n. 42

Fessenden, Reginald, 1, 63

Film industry: early history, 2, 34; and growing size of theaters, 56; profits: —in 1930s, 48, 55, 119, 121, 123; —in 1940s, 240n. 48; reliance on live music, 34–37, 88–89; and theater ownership, 227n. 28; transition to sound, 47–50, 214nn. 29–32. *See also* Employment patterns; Strikes; Unemployment among musicians; Wages; Working conditions

Fischer, Phil, 179–80, 198

Fleming, Bob, 104, 167, 190

Fly, James L., 135, 143–45, 232n. 17

Forbstein, Leo, 92

Foreign orchestras, 29–30, 211nn. 46–47

Frequency modulation (FM): in collective bargaining, 165–66, 177–78, 180, 187; in congressional debates, 169–70; developed, 132, 165; FCC reverses policy, 165

Gabler, Milton, 156

Galbraith, John Kenneth, 81

Gamble, Thomas, 128–29

Gardner, Samuel, 61

Gillette, J. W., 90

Gold, Ernest, 94–95

Gompers, Samuel, 22, 24

Goodman, Benny, 80, 134

Government: antitrust laws, 112–13; early radio, 77, 83; and employer associations, 80; and Great Depression, 117–19; ideology, 173, 199; impact of congressional elections, 145–46, 163; injunctions against labor, 52–53; legislators criticize Petrillo, 169–70, 182; NAPA, 86–87; proper role of, 3, 199–200; and radio ownership, 171–72, 237nn. 18–19; and record ban of 1942–44, 141, 155, 157–58

Great Depression, 71, 85, 117–18, 119, 123

Griffith, D. W., 94, 223n. 16

Gruen, Henry, 98, 128

Gurney, Chad, 171

Hartley, Fred A., 174, 181, 182

Hayden, A. C., 56, 83

Heifetz, Jascha, 156

Hendrickson, Al, 91, 93

Herman, Woody, 80, 129

Hierarchy in the workplace: sidemen and leaders, 11; in studios, 91, 94, 99; and technological change, 194, 200; in theaters, 45–46

Hild, Oscar, 129

Hilman, Roc, 167, 190

Hinton, Milt, 76

Hiring process: leader's role, 11–12; in studios, 91, 222n. 11, 224n. 27; in theaters, 40

Hitchcock, Bill, 138–39, 232n. 18

Hokanson, Nels, 9

Holifield, Chet, 170

Hoover, Herbert, 63, 211n. 47, 217n. 11

Hubbard, Stanley E., 141

Humphreys, Dorothy S., 64–65

Identity: artist vs. worker controversy, 22–24, 26, 28; black musicians, 31, corporate psyche, 111–12, social alienation, 194; in studios, 100

Ideology: communism, 172; irreconcilable positions, 154; labor's point of view, 114–15, 185; lawmakers' view, 173, 199; management's view, 111–12, 123. *See also* Language and labor

Independent Radio Network Affiliates (IRNA), 124, 228n. 38

Interlochen incident, 141–43

International Alliance of Theatrical and Stage Employees (IATSE), 175

Jensen, Peter L., 1

Johnson, Lyndon B., 171

Joplin, Scott, 8

Jukeboxes: developed, 78, 225n. 2, 229n. 55; and first record ban, 147; impact on employment, 78, 107, 130; in late 1930s, 108, 130, *131,* 232–33n. 20; profits from, 153, 240n. 48; "telephone" jukebox, 132. *See also* Record industry

Library of Congress Cataloging-in-Publication Data

Kraft, James P.
  Stage to studio : musicians and the sound revolution, 1890–1950 /
  James P. Kraft.
      p.   cm. — (Studies in industry and society ; 9)
    Includes bibliographical references and index.
    ISBN 0-8018-5089-4 (alk. paper)
      1. Musicians—Employment—United States.   2. Musicians
  —Effect of technological innovations on—United States.
  3. Musicians—Legal status, laws, etc.—United States.   4. Industrial
  relations—United States.   5. Labor movement—United States.
  I. Title.   II. Series.
  ML3795.K82   1996
  331'.04178'0973—dc20   95-43923 .